UNIT 5 Probability

UNIT 6 Formulae

UNIT 7 Mean, Median, and Mode

UNIT 8 Borrowing and Saving

UNIT 9 Pictorial Representation

UNIT 10 Rates and Taxes

UNIT 11 Index Numbers

UNIT 12 Estimates and Errors

ANSWERS

21117090L   TS

Modular Mathematics

Certificate Arithmetic

Modular Mathematics Organization

Heinemann Educational Books
London and Edinburgh

H·E·B

CONTENTS

UNIT 1 Number Skills

A Whole numbers

In the following exercise the signs $+$ and $-$ have been omitted. Choose the appropriate signs to put into the spaces so that correct statements are made.

Example

$$17 \quad 4 \quad 7 = 20$$
Answer: $17 - 4 + 7 = 20$

Exercise

Copy and complete:

1 13   4   3 = 20   **2** 20   6   4 = 10

3 19   3   8 = 24   **4** 30   15   5 = 20

5 14   11   6 = 19   **6** 27   17   16 = 28

Example

Calculate $15 - 2 \times (5 - 3)$.
We follow this sequence:

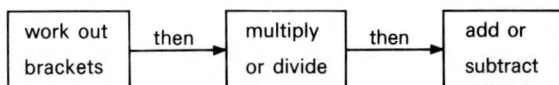

work out brackets	then	multiply or divide	then	add or subtract

$$15 - 2 \times (5 - 3)$$
$$= 15 - 2 \times 2$$
$$= 15 - 4$$
$$= 11$$

Exercise

Do the following, setting out working as above:

7 $5 \times (7 + 8) + 4$ **8** $5 \times (8 - 7) + 4$

9 $5 + 8 \times (7 - 4)$ **10** $(5 + 8) \times (7 - 4)$

11 $8 + (5 + 7) \div 4$ **12** $(8 + 4) \div (5 + 7)$

13 $8 - 4 \times (7 - 5)$ **14** $(5 + 7) \div (8 - 4)$

Continue with Section B

B Multiplication and division by 10, 100, 1000

When we multiply 2.8 by 10 we do the following:

Tens	Units . tenths
	2 . 8
2	8 . 0

$2.8 \times 10 = 28$ $\times 10$

To multiply by 10 we move the figures one place to the left.

When we multiply 2.8 by 100 we do the following:

Hundreds	Tens	Units . tenths
		2 . 8
2	8	0 . 0

$2.8 \times 100 = 280$ $\times 100$

To multiply by 100 we move the figures two places to the left.

Exercise

Copy and complete:

1 $3.7 \times 10 =$	**2** $12.4 \times 10 =$	**3** $7.24 \times 10 =$
4 $3.7 \times 100 =$	**5** $12.4 \times 100 =$	**6** $7.24 \times 100 =$
7 $3.7 \times 1000 =$	**8** $12.4 \times 1000 =$	**9** $7.24 \times 1000 =$

Example

Calculate (a) 3.2×20 and (b) 6.28×200

(a) $3.2 \times 20 = 3.2 \times 2 \times 10$
$= 6.4 \times 10$
$= 64$

(b) $6.28 \times 200 = 6.28 \times 2 \times 100$
$= 12.56 \times 100$
$= 1256$

Exercise

Calculate the following, setting out your working as above:

10 4.6×20	**11** 5.17×30	**12** 3.82×40
13 4.6×200	**14** 5.17×300	**15** 3.82×400
16 4.6×2000	**17** 5.17×3000	**18** 3.82×4000
19 0.01×300	**20** 0.2×200	**21** 0.03×20

When we divide 2.8 by 10 we do the following:

Units . tenths	hundredths
2 . 8	
0 . 2	8

$2.8 \div 10 = 0.28$ $\div 10$

To divide by 10 we move the figures one place to the right.

When we divide 2.8 by 100 we do the following:

Units . tenths	hundredths	thousandths
2 . 8		
0 . 0	→2	→8

$2.8 \div 100 = 0.028$

$\div 100$

To divide by 100 we move the figures two places to the right.

Exercise

Copy and complete the following:

22 $22.6 \div 10 =$	**23** $13.4 \div 10 =$	**24** $5.1 \div 10 =$
25 $22.6 \div 100 =$	**26** $13.4 \div 100 =$	**27** $5.1 \div 100 =$
28 $22.6 \div 1000 =$	**29** $13.4 \div 1000 =$	**30** $5.1 \div 1000 =$

Continue with Section C

C Multiplication and division of decimals

Calculate 4.1×5.2

$$
\begin{array}{r}
4.1 \\
\times\ 5.2 \\
\hline
8\,2 \\
2\,0\,5\,0 \\
\hline
2\,1.3\,2 \\
\end{array}
$$

point placed here.

Rule

The number of decimal places in the answer is equal to the total number of decimal places in the numbers being multiplied.

Example

Calculate 0.23×0.3

0.23 ←——— 2 decimal places
$\times 0.3$ ←——— 1 decimal place
0.069 ←——— 3 decimal places

Exercise

Calculate the following:

1 3.4×2.3	**2** 64.1×2.3	**3** 17.6×0.41
4 1.23×3.6	**5** $0.2 \times 0.2 \times 0.6$	**6** $0.4 \times 0.4 \times 0.4$
7 3.05×0.4	**8** 14.5×24	**9** $0.2 \times 0.3 \times 0.4$

Example

Calculate $76.8 \div 3.2$

Before dividing we make the number we are dividing by a whole number.

$$\frac{76.8}{3.2} = \frac{76.8 \times 10}{3.2 \times 10} = \frac{768}{32}$$

$$76.8 \div 3.2 = 24$$

Working

```
      24
32 ) 768
     64↓
     128
     128
       0
```

Example

Calculate $518 \div 3.7$

$$\frac{518.}{3.7} = \frac{518. \times 10}{3.7 \times 10} = \frac{5180}{37}$$

$$518 \div 3.7 = 140$$

Working

```
       140
37 ) 5180
     37↓|
     148|
     148↓
       0
```

Exercise

Calculate the following:

10 $22.8 \div 3.8$	**11** $782 \div 3.4$	**12** $55.2 \div 2.4$
13 $13.44 \div 0.42$	**14** $9.1 \div 0.26$	**15** $17.92 \div 0.64$
16 $40 \div 0.02$	**17** $2 \div 0.005$	**18** $8 \div 0.4$

Example

Calculate $4.86 \div 1.5$

$$\frac{4.86}{1.5} = \frac{4.86 \times 10}{1.5 \times 10} = \frac{48.6}{15}$$

$$4.86 \div 1.5 = 3.24$$

Working

```
        3.24
15 ) 48.60
     45 ↓|
      3 6|
      3 0↓
        60
        60
         0
```

Exercise

Calculate the following:

19 $5.72 \div 5.5$	**20** $4.14 \div 1.2$	**21** $15.08 \div 6.5$

So far the divisions have given exact answers but we may have to do a division like this one:

$$2.93 \div 1.3$$

$$\frac{2.93}{1.3} = \frac{29.3}{13}$$

```
          2.253 . . .
13 ) 29.3000
     26 ↓|||
      3 3|||
      2 6↓||
        70|
        65↓
        50
        39↓
       110
```

In a division like this, we will stop when we reach a given number of decimal places.

Example

Divide 29.3 by 13 and give answer correct to 1 decimal place (1 dp).
The answer is wanted to 1 decimal place, so we take the division to 2 decimal places.

$$29.3 \div 13 = 2.25$$

$$
\begin{array}{r}
2.25\ldots \\
13\,)\,\overline{29.30} \\
26\downarrow| \\
\overline{3\,3} \\
2\,6\downarrow \\
\overline{70} \\
65 \\
\end{array}
$$

Look at the figure in the second decimal place.
If this is 5 or more we write the answer as 2.3.
If this is less than 5 write the answer as 2.2.

$$29.3 \div 13 = 2.3 \text{ (to 1 dp)}$$

Exercise

Calculate the following to 1 decimal place:

22 $17.4 \div 11$ **23** $43.1 \div 15$ **24** $24.3 \div 21$

Example

Divide 29.3 by 13 and give the answer correct to 2 decimal places

$$29.3 \div 13 = 2.253\ldots$$

$$
\begin{array}{r}
2.253\ldots \\
13\,)\,\overline{29.300} \\
26 \\
\overline{3\,3} \\
2\,6 \\
\overline{70} \\
65 \\
\overline{50} \\
\end{array}
$$

Look at the figure in the third decimal place.
If this is 5 or more write answer as 2.26.
If this is less than 5 write answer as 2.25.

$$29.3 \div 13 = 2.25 \text{ (to 2 dp)}$$

Exercise

Calculate the following to 2 decimal places:

25 $2.54 \div 23$ **26** $3.17 \div 15$ **27** $5.23 \div 31$

Continue with Section D

D Decimal calculations

Example

Calculate $13.71 - 15.38 + 12.9$

$$13.71\overparen{-15.38 + 12.9}$$
$$= 13.71 + 12.9 - 15.38$$
$$= 26.61 - 15.38$$
$$= 11.23$$

> Change order so that addition is done first.

$$\begin{array}{r} 13.71 \\ +12.9 \\ \hline 26.61 \end{array} \qquad \begin{array}{r} 26.61 \\ -15.38 \\ \hline 11.23 \end{array}$$

Exercise

Calculate the following:

1 $15.4 + 1.32 + 16.3$

2 $2.7 + 31.7 - 0.63$

3 $16.5 - 21.1 + 13.4$

4 $3.13 - 0.84 + 2.6$

5 $4.23 + 16.7 - 0.8 - 6.35$

Example

Calculate $1.5 + 2 \times (2.5 - 0.3)$.
Follow this sequence:

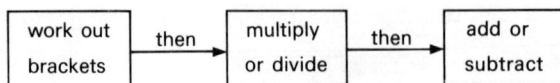

work out brackets	then	multiply or divide	then	add or subtract

$$1.5 + 2 \times \underbrace{(2.5 - 0.3)}$$
$$= 1.5 + 2 \times 2.2$$
$$= 1.5 + 4.4$$
$$= 5.9$$

Exercise

Calculate the following, setting out working as above:

6 $5 \times (0.7 + 0.8) + 4.2$

7 $5 \times (0.8 - 0.7) + 4.2$

8 $(0.8 + 4.2) \times (0.7 - 0.2)$

9 $4.2 + 2 \times (0.8 - 0.5)$

10 $0.8 + (4.2 + 0.7) \div 0.7$

Continue with Section E

E Equivalent fractions

Example

Express $\frac{12}{16}$ as a fraction in lowest terms. $\frac{12}{16} = \frac{3}{4}$ (÷4)

Exercise

Express the following fractions in their lowest terms:

1 $\frac{15}{20}$ **2** $\frac{6}{9}$ **3** $\frac{27}{30}$ **4** $\frac{12}{18}$ **5** $\frac{20}{30}$ **6** $\frac{12}{16}$

Example

Express the following fraction with the given denominator.

$$\frac{5}{8} = \frac{\blacksquare}{24}$$

$$\frac{5}{8} = \frac{15}{24} \quad (\times 3)$$

Exercise

Express the following fractions with the given denominators:

7 $\frac{5}{6} = \frac{\blacksquare}{36}$ **8** $\frac{3}{8} = \frac{\blacksquare}{32}$ **9** $\frac{4}{5} = \frac{\blacksquare}{30}$

10 $\frac{3}{7} = \frac{\blacksquare}{49}$ **11** $\frac{2}{9} = \frac{\blacksquare}{36}$ **12** $\frac{2}{25} = \frac{\blacksquare}{75}$

Example

Which of the fractions $\frac{4}{5}$ and $\frac{5}{8}$ is the greater?

To compare fractions we must bring them to the same name – in this case we make the denominator 40.

$$\frac{4}{5} = \frac{32}{40}; \quad \frac{5}{8} = \frac{25}{40}$$

$\frac{32}{40}$ is greater than $\frac{25}{40}$ so $\frac{4}{5}$ is greater than $\frac{5}{8}$.

We write $\frac{32}{40} > \frac{25}{40}$ so $\frac{4}{5} > \frac{5}{8}$.

Exercise

In each of the following find which fraction is the greater:

13 $\frac{3}{4}, \frac{5}{8}$ **14** $\frac{2}{5}, \frac{3}{10}$ **15** $\frac{2}{3}, \frac{7}{9}$

16 $\frac{2}{5}, \frac{1}{3}$ **17** $\frac{5}{6}, \frac{7}{8}$ **18** $\frac{3}{4}, \frac{5}{6}$

Continue with Section F

F Addition and subtraction of fractions

Example

Calculate (a) $\frac{1}{3}+\frac{4}{9}$ and (b) $\frac{2}{5}+\frac{3}{4}$.

$$
\begin{aligned}
\text{(a)} \quad & \frac{1}{3}+\frac{4}{9} \\
= & \frac{3}{9}+\frac{4}{9} \\
= & \frac{7}{9}
\end{aligned}
\qquad
\begin{aligned}
\text{(b)} \quad & \frac{2}{5}+\frac{3}{4} \\
= & \frac{8}{20}+\frac{15}{20} \\
= & \frac{23}{20} \\
= & \frac{20}{20}+\frac{3}{20} \\
= & 1\frac{3}{20}
\end{aligned}
$$

Exercise

Calculate:

1 $\frac{1}{4}+\frac{1}{6}$ **2** $\frac{1}{4}+\frac{5}{12}$ **3** $\frac{1}{3}+\frac{2}{7}$

4 $\frac{2}{3}+\frac{5}{6}$ **5** $\frac{1}{2}+\frac{2}{3}$ **6** $\frac{4}{5}+\frac{1}{3}$

Example

Calculate (a) $\frac{5}{6}-\frac{1}{3}$ and (b) $\frac{7}{12}-\frac{1}{5}$.

$$
\begin{aligned}
\text{(a)} \quad & \frac{5}{6}-\frac{1}{3} \\
= & \frac{5}{6}-\frac{2}{6} \\
= & \frac{3}{6} \\
= & \frac{1}{2}
\end{aligned}
\qquad
\begin{aligned}
\text{(b)} \quad & \frac{7}{12}-\frac{1}{5} \\
= & \frac{35}{60}-\frac{12}{60} \\
= & \frac{23}{60}
\end{aligned}
$$

Exercise

Calculate:

7 $\frac{5}{12}-\frac{1}{3}$ **8** $\frac{5}{8}-\frac{1}{4}$ **9** $\frac{7}{9}-\frac{1}{3}$

10 $\frac{4}{5}-\frac{1}{2}$ **11** $\frac{7}{8}-\frac{1}{3}$ **12** $\frac{2}{3}-\frac{1}{5}$

Continue with Section G

G Multiplication of fractions

Example

What is $\frac{3}{4}$ of 8?

$$
\begin{aligned}
\frac{1}{4} \text{ of } 8 &= 2 \\
\frac{3}{4} \text{ of } 8 &= 3 \times 2 = 6
\end{aligned}
$$

Exercise

Calculate:

1 $\frac{1}{2}$ of 16   **2** $\frac{1}{5}$ of 25   **3** $\frac{1}{3}$ of 24

4 $\frac{3}{5}$ of 15   **5** $\frac{2}{3}$ of 24   **6** $\frac{7}{8}$ of 24

Example

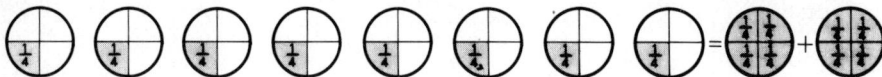

$$8 \times \tfrac{1}{4} = \quad \tfrac{1}{4} + \tfrac{1}{4} + \tfrac{1}{4} + \tfrac{1}{4} + \tfrac{1}{4} + \tfrac{1}{4} + \tfrac{1}{4} + \tfrac{1}{4}$$
$$= \tfrac{8}{4}$$
$$= 2$$

Example

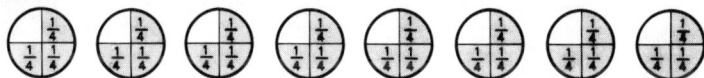

$$8 \times \tfrac{3}{4} = \quad \tfrac{3}{4} + \tfrac{3}{4} + \tfrac{3}{4} + \tfrac{3}{4} + \tfrac{3}{4} + \tfrac{3}{4} + \tfrac{3}{4} + \tfrac{3}{4}$$
$$= \tfrac{24}{4}$$

$$= 6$$

Example

Calculate $\frac{3}{4} \times 16$.

$$\tfrac{3}{4} \times 16 = \tfrac{48}{4} = 12$$

Exercise

Calculate:

7 $9 \times \frac{1}{3}$ **8** $15 \times \frac{1}{5}$ **9** $18 \times \frac{1}{6}$

10 $9 \times \frac{2}{3}$ **11** $15 \times \frac{3}{5}$ **12** $18 \times \frac{5}{6}$

13 $\frac{1}{4} \times 12$ **14** $\frac{1}{5} \times 20$ **15** $\frac{1}{6} \times 24$

16 $\frac{3}{4} \times 12$ **17** $\frac{3}{5} \times 20$ **18** $\frac{5}{6} \times 24$

Example

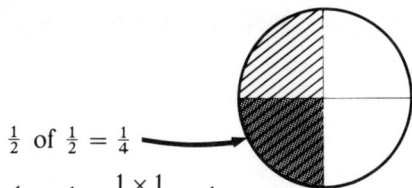

$$\tfrac{1}{2} \text{ of } \tfrac{1}{2} = \tfrac{1}{4}$$
$$\tfrac{1}{2} \times \tfrac{1}{2} = \frac{1 \times 1}{2 \times 2} = \tfrac{1}{4}$$

So, $\frac{1}{2}$ of $\frac{1}{2} = \frac{1}{2} \times \frac{1}{2}$

Example

$$\frac{2}{3} \times \frac{3}{4}$$

$$= \frac{2 \times 3}{3 \times 4}$$

$$= \frac{6}{12}$$

$$= \frac{1}{2}$$

Exercise

Calculate:

19 $\frac{1}{3}$ of $\frac{1}{3}$	**20** $\frac{1}{2}$ of $\frac{1}{4}$	**21** $\frac{1}{4}$ of $\frac{1}{2}$
22 $\frac{1}{2} \times \frac{3}{4}$	**23** $\frac{1}{4} \times \frac{2}{3}$	**24** $\frac{1}{5} \times \frac{3}{5}$
25 $\frac{2}{3} \times \frac{3}{5}$	**26** $\frac{2}{5} \times \frac{2}{3}$	**27** $\frac{3}{4} \times \frac{2}{5}$

Continue with Section H

H Progress check

Exercise

Calculate the following:

1 $14 + 9 - 6$ **2** $23 - 15 + 12$

3 $12 - 3 \times (7 - 4)$ **4** $(9 + 5) \div (8 - 6)$

5   (a) 13.7×10 (b) 13.7×100 (c) 13.7×1000

6   (a) 4.16×20 (b) 4.16×200 (c) 4.16×2000

7   (a) $24.3 \div 10$ (b) $24.3 \div 100$ (c) $24.3 \div 1000$

8 $0.3 \times 0.3 \times 0.3$ **9** 3.14×1.6

10 15.2×0.23 **11** $5.52 \div 2.3$

12 $9.72 \div 0.36$ **13** $9.72 \div 1.5$ (correct to 1 dp)

14 $4.13 - 1.6 + 15.87$ **15** $6 \times (0.3 + 1.2) + 4.3$

16   Bring to lowest terms: (a) $\dfrac{15}{25}$ (b) $\dfrac{21}{28}$ (c) $\dfrac{50}{100}$

17   Express the following fractions with the given denominators:

(a) $\dfrac{2}{3} = \dfrac{\blacksquare}{18}$ (b) $\dfrac{4}{5} = \dfrac{\blacksquare}{10}$ (c) $\dfrac{5}{6} = \dfrac{\blacksquare}{30}$

18   Which of the following two fractions is the greater? $\frac{4}{5}$ or $\frac{7}{10}$

19   Calculate: (a) $\frac{2}{6} + \frac{1}{10}$ (b) $\frac{9}{7} + \frac{3}{5}$

20   Calculate: (a) $\frac{9}{10} - \frac{2}{5}$ (b) $\frac{4}{5} - \frac{1}{2}$

21   Calculate: (a) $\frac{1}{3}$ of 27   (b) $16 \times \frac{3}{4}$ (c) $\frac{3}{8} \times 24$

22   Calculate: (a) $\frac{1}{4}$ of $\frac{1}{8}$ (b) $\frac{1}{3} \times \frac{2}{5}$ (c) $\frac{3}{5} \times \frac{5}{8}$

Ask your teacher what to do next

I Mixed numbers

Example

$$1\tfrac{2}{5} + 2\tfrac{3}{4}$$
$$= 3 + \tfrac{2}{5} + \tfrac{3}{4}$$
$$= 3 + \tfrac{8}{20} + \tfrac{15}{20}$$
$$= 3 + \tfrac{23}{20}$$
$$= 3 + 1 + \tfrac{3}{20}$$
$$= 4 + \tfrac{3}{20}$$
$$= 4\tfrac{3}{20}$$

$$3\tfrac{1}{2} - 1\tfrac{5}{8}$$
$$= 2\tfrac{1}{2} - \tfrac{5}{8}$$
$$= 2 + \tfrac{4}{8} - \tfrac{5}{8}$$
$$= 1 + \tfrac{12}{8} - \tfrac{5}{8}$$
$$= 1\tfrac{7}{8}$$

Exercise

Calculate:

1 $3\tfrac{1}{6} + 1\tfrac{2}{3}$

2 $2\tfrac{1}{2} + 1\tfrac{5}{6}$

3 $\tfrac{8}{9} + \tfrac{2}{3} + \tfrac{1}{6}$

4 $3\tfrac{1}{2} - 1\tfrac{3}{4}$

5 $3\tfrac{1}{8} - 2\tfrac{1}{5}$

6 $14\tfrac{2}{3} - 2\tfrac{7}{8}$

7 $\tfrac{1}{2} + \tfrac{2}{3} - \tfrac{3}{4}$

8 $3\tfrac{1}{2} + 1\tfrac{1}{4} - 1\tfrac{1}{8}$

9 $3\tfrac{2}{5} + 1\tfrac{1}{10} - 2\tfrac{1}{3}$

Example

$$1\tfrac{3}{5} \times 3\tfrac{1}{4}$$
$$= \frac{8}{5} \times \frac{13}{4}$$
$$= \frac{{}^{2}\cancel{8} \times 13}{5 \times \cancel{4}_{1}} \quad \left[\begin{array}{l} \text{Dividing above and} \\ \text{below by 4.} \end{array} \right]$$
$$= \frac{2 \times 13}{5 \times 1}$$
$$= \tfrac{26}{5}$$
$$= 5\tfrac{1}{5}$$

Exercise

Calculate:

10 $1\tfrac{1}{2} \times 3\tfrac{1}{3}$

11 $4\tfrac{1}{3} \times 1\tfrac{1}{5}$

12 $2\tfrac{1}{4} \times 1\tfrac{1}{3}$

13 $2\tfrac{1}{2} \times 3\tfrac{1}{3}$

14 $2\tfrac{1}{5} \times 3\tfrac{3}{4}$

15 $1\tfrac{2}{3} \times 1\tfrac{1}{5}$

Continue with Section J

J Division of fractions

$$6 \div \tfrac{2}{3} = \frac{6}{\tfrac{2}{3}}$$

$$= \frac{6 \times \tfrac{3}{2}}{\tfrac{2}{3} \times \tfrac{3}{2}}$$

$$\left[\begin{array}{l}\text{Multiplying numerator and}\\ \text{denominator by } \tfrac{3}{2} \text{ so that}\\ \text{denominator becomes 1.}\end{array}\right]$$

$$= 6 \times \tfrac{3}{2}$$

$$= 9$$

So dividing by $\tfrac{2}{3}$ is the same as multiplying by $\tfrac{3}{2}$.

Example

Calculate $\tfrac{3}{8} \div \tfrac{1}{2}$.

$$\tfrac{3}{8} \div \tfrac{1}{2} = \tfrac{3}{8} \times \tfrac{2}{1}$$

$$= \frac{3 \times 2^1}{{}_4 8 \times 1} \quad \left[\begin{array}{l}\text{Dividing above and}\\ \text{below by 2.}\end{array}\right]$$

$$= \frac{3 \times 1}{4 \times 1}$$

$$= \tfrac{3}{4}$$

Exercise

Calculate:

1 $\tfrac{4}{9} \div \tfrac{1}{3}$

2 $\tfrac{2}{3} \div \tfrac{5}{6}$

3 $\tfrac{3}{8} \div \tfrac{1}{4}$

Example

Calculate $3\tfrac{1}{4} \div 1\tfrac{1}{2}$.

$$3\tfrac{1}{4} \div 1\tfrac{1}{2} = \tfrac{13}{4} \div \tfrac{3}{2}$$

$$= \tfrac{13}{4} \times \tfrac{2}{3}$$

$$= \frac{13 \times 2^1}{{}_2 4 \times 3} \quad \left[\begin{array}{l}\text{Dividing above and}\\ \text{below by 2.}\end{array}\right]$$

$$= \frac{13 \times 1}{2 \times 3}$$

$$= \tfrac{13}{6}$$

$$= 2\tfrac{1}{6}$$

Exercise

Calculate:

4 $2\tfrac{1}{2} \div \tfrac{3}{4}$

5 $2\tfrac{3}{4} \div 1\tfrac{1}{2}$

6 $3\tfrac{1}{3} \div 1\tfrac{1}{3}$

Continue with Section K

K Conversion to decimal fractions

Example

Express $\frac{2}{5}$ as a decimal fraction.
We try to express the fraction with a denominator of 10, 100, 1000, ...

$$\frac{2}{5} = \frac{4}{10} = 0.4 \quad (\times 2)$$

Example

Express $\frac{1}{20}$ as a decimal fraction.

$$\frac{1}{20} = \frac{5}{100} = 0.05 \quad (\times 5)$$

Exercise

Express the following fractions as decimal fractions:

1 $\frac{1}{2}$ 　　　　　　　　 2 $\frac{4}{5}$ 　　　　　　　　 3 $\frac{3}{20}$

4 $\frac{1}{25}$ 　　　　　　　　 5 $\frac{1}{50}$ 　　　　　　　　 6 $\frac{3}{4}$

Example

Convert $\frac{2}{7}$ to a decimal fraction, giving the answer correct to 2 decimal places.

$\frac{2}{7}$ means the same as $2 \div 7$

$$\begin{array}{r} 0.285\ldots \\ 7\overline{)2.000} \\ \underline{1\,4} \\ 60 \\ \underline{56} \\ 40 \end{array}$$

$\frac{2}{7} = 0.285\ldots$

So $\frac{2}{7} = 0.29$ (to 2 dp)

Exercise

Convert the following common fractions to decimal fractions, giving answers correct to 2 decimal places:

7 $\frac{7}{8}$ 　　　　　　　　 8 $\frac{2}{9}$ 　　　　　　　　 9 $\frac{3}{11}$

10 $\frac{3}{16}$ 　　　　　　　　 11 $\frac{5}{32}$ 　　　　　　　　 12 $\frac{5}{6}$

Continue with Section L

L Conversion to percentages

Example

Express $\frac{2}{5}$ as a percentage.

$$\frac{2}{5} = 0.40$$
$$= \frac{40}{100}$$
$$= 40\% \qquad \left[\begin{array}{l} \text{'per cent' means} \\ \text{'out of a hundred'} \end{array} \right]$$

Example

Express $\frac{7}{20}$ as a percentage.

$$\frac{7}{20} = 0.35$$
$$= \frac{35}{100}$$
$$= 35\%$$

Exercise

Express the following fractions as percentages:

1 $\frac{1}{2}$ **2** $\frac{4}{5}$ **3** $\frac{3}{20}$ **4** $\frac{1}{25}$ **5** $\frac{1}{50}$ **6** $\frac{3}{4}$

Example

Express $\frac{7}{12}$ as a percentage correct to 1 decimal place.

$\frac{7}{12}$ means the same as $7 \div 12$

$$
\begin{array}{r}
0.5833\ldots \\
12\,)\overline{7.0000} \\
\underline{6\,0\downarrow} \\
1\,00 \\
\underline{96\downarrow} \\
40 \\
\underline{36\downarrow} \\
40
\end{array}
$$

$$\frac{7}{12} = 0.5833\ldots$$
$$= \frac{58.33\ldots}{100}$$
$$= 58.33\ldots\%$$
$$= 58.3\% \text{ (to 1 dp)}$$

Exercise

Express the following vulgar fractions as percentages correct to 1 decimal place:

7 $\frac{3}{8}$ **8** $\frac{4}{9}$ **9** $\frac{2}{11}$ **10** $\frac{5}{16}$ **11** $\frac{7}{32}$ **12** $\frac{1}{3}$

13 Copy and complete the following table which shows three ways of expressing the same quantity.

Vulgar fraction	Decimal fraction	Percentage
$\frac{1}{2}$	0.5	50%
$\frac{1}{5}$		
$\frac{3}{5}$		
$\frac{4}{9}$	0.444 ...	44.4%
$\frac{1}{3}$		
$\frac{2}{3}$		

Example

Express 24% as (a) a decimal fraction and (b) a vulgar fraction in lowest terms.

(a) $24\% = 0.24$

(b) $0.24 = \frac{24}{100} = \frac{6}{25}$

$$\overset{\div 4}{\underset{\div 4}{\frown}}$$

Example

Express $2\frac{1}{2}\%$ as (a) a decimal fraction and (b) a vulgar fraction in lowest terms.

(a) $2\frac{1}{2}\% = 0.025$

(b) $0.025 = \frac{25}{1000} = \frac{5}{200} = \frac{1}{40}$

$$\overset{\div 5}{\underset{\div 5}{\frown}}\overset{\div 5}{\underset{\div 5}{\frown}}$$

Exercise

Express the following percentages as (a) decimal fractions and (b) vulgar fractions in lowest terms:

14 12%   **15** 25%   **16** 40%

17 8%   **18** $12\frac{1}{2}\%$ **19** $7\frac{1}{2}\%$

Continue with Section M

M Calculation of square roots

We can use square root tables to find the square roots of numbers between 1 and 100. To find the square roots of numbers less than 1 or greater than 100, the numbers involved must first of all be factorized in a special way.

Let us consider numbers greater than 100.

You should know that

$$\sqrt{100} = 10$$
$$\sqrt{10\,000} = 100$$
$$\sqrt{1\,000\,000} = 1000$$

and we make use of these square roots in finding the square roots of other numbers.

Example

Find $\sqrt{6400}$.

$$\sqrt{6400} = \sqrt{64 \times 100}$$
$$= \sqrt{64} \times \sqrt{100}$$
$$= 8 \times 10$$
$$= 80$$

Example

Find $\sqrt{90\,000}$.

$$\sqrt{90\,000} = \sqrt{9 \times 10\,000}$$
$$= 3 \times 100$$
$$= 300$$

Exercise

Calculate the following:

1 $\sqrt{3600}$ 2 $\sqrt{4900}$ 3 $\sqrt{8100}$

4 $\sqrt{40\,000}$ 5 $\sqrt{160\,000}$ 6 $\sqrt{250\,000}$

For numbers less than 1 we use these same square roots but divide by them instead of multiplying by them.

Example

Find (a) $\sqrt{0.09}$ and (b) $\sqrt{0.0036}$.

(a) $\sqrt{0.09} = \sqrt{\dfrac{9}{100}}$

$\quad\quad = \dfrac{\sqrt{9}}{\sqrt{100}}$

$\quad\quad = \dfrac{3}{10}$

$\quad\quad = 0.3$

(b) $\sqrt{0.0036} = \sqrt{\dfrac{36}{10\,000}}$

$\quad\quad\quad = \dfrac{\sqrt{36}}{\sqrt{10\,000}}$

$\quad\quad\quad = \dfrac{6}{100}$

$\quad\quad\quad = 0.06$

Exercise

Calculate the following:

7 $\sqrt{0.16}$

8 $\sqrt{0.0049}$

9 $\sqrt{0.0144}$

10 $\sqrt{0.000\,004}$

11 $\sqrt{0.000\,169}$

12 $\sqrt{0.0196}$

Continue with Section N

N Estimation of square roots

Example

Which of the following is nearest in value to $\sqrt{1740}$?
A. 12 B. 40 C. 120 D. 400 E. 1200

Try A $12 \times 12 = 144$ ◄——too small
    B $40 \times 40 = \boxed{1600}$
    C $120 \times 120 = 14\,400$ ◄——too large

D and E would give even larger values so we can ignore them.

B gives the nearest answer.

Example

Which of the following is nearest in value to $\sqrt{0.0142}$?
A. 0.12 B. 0.4 C. 1.2 D. 4 E. 12

Try A $0.12 \times 0.12 = \boxed{0.0144}$ which is nearly 0.0142
    B $0.4 \times 0.4 = 0.16$ ◄——too large

C, D, and E would give even larger values so we can ignore them.

A gives the nearest answer.

Exercise

1 Which of the following is nearest in value to $\sqrt{6510}$?
   A. 0.8   B. 2.5   C. 8   D. 25   E. 80

2 Which of the following is nearest in value to $\sqrt{14\,900}$?
   A. 1.2   B. 4   C. 12   D. 40   E. 120

3 Which of the following is nearest in value to $\sqrt{0.0052}$?
   A. 0.007   B. 0.02   C. 0.07   D. 0.2   E. 0.7

4 Which of the following is nearest in value to $\sqrt{0.83}$?
   A. 0.03   B. 0.09   C. 0.3   D. 0.9   E. 3

Continue with Section O

⊙ Estimation of cube roots

$2^3 = 2 \times 2 \times 2 = 8$

We say the **cube root** of 8 is 2 and write this as $\sqrt[3]{8} = 2$.

$3^3 = 3 \times 3 \times 3 = 27$ so $\sqrt[3]{27} = 3$

$4^3 = 4 \times 4 \times 4 = 64$ so $\sqrt[3]{64} = 4$

Also $0.2^3 = 0.2 \times 0.2 \times 0.2 = 0.008$ so $\sqrt[3]{0.008} = 0.2$

$0.3^3 = 0.3 \times 0.3 \times 0.3 = 0.027$ so $\sqrt[3]{0.027} = 0.3$

Example

Which of the following is nearest in value to $\sqrt[3]{0.007\,67}$?
A. 0.2   B. 0.02   C. 0.002   D. 0.0002   E. 0.000 02

Try A $0.2 \times 0.2 \times 0.2 =$ ⎡0.008⎤ which is nearly 0.007 67

B $0.02 \times 0.02 \times 0.02 = 0.000\,008$ ⟵ too small

C, D and E would give even smaller values so we can ignore them.

A is nearest in value.

Example

Which of the following is nearest in value to $\sqrt[3]{30\,000}$?
A. 3   B. 30   C. 300   D. 3000   E. 30 000

Try A $3^3 = 3 \times 3 \times 3 = 27$ ⟵ too small

B $30^3 = 30 \times 30 \times 30 =$ ⎡27 000⎤ which is nearer 30 000 than the others

C $300^3 = 300 \times 300 \times 300 = 27\,000\,000$ ⟵ too large

D and E would give even larger values so we can ignore them.

B is nearest in value.

Exercise

1 Which of the following is nearest in value to $\sqrt[3]{0.009\,12}$?

A. 0.2   B. 0.02   C. 0.002   D. 0.0002   E. 0.000 02

2 Which of the following is nearest in value to $\sqrt[3]{0.000\,031}$?

A. 0.3   B. 0.03   C. 0.003   D. 0.0003   E. 0.000 03

3 Which of the following is nearest in value to $\sqrt[3]{30\,000\,000}$?

A. 3   B. 30   C. 300   D. 3000   E. 30 000

4 Which of the following is nearest in value to $\sqrt[3]{59\,000}$?

A. 4   B. 40   C. 400   D. 4000   E. 40 000

Continue with Section P

P Progress check

Exercise

Calculate the following:

1 $2\frac{3}{5}+1\frac{1}{3}$ **2** $3\frac{2}{3}-1\frac{5}{6}$

3 $1\frac{1}{2}+2\frac{3}{4}-1\frac{1}{3}$ **4** $2\frac{1}{3}\times2\frac{1}{4}$

5 $3\frac{1}{3}+1\frac{1}{5}$

6 Express the following as decimal fractions:

  (a) $\frac{3}{5}$ (b) $\frac{3}{10}$ (c) $\frac{1}{4}$

7 Express the following as percentages:

  (a) $\frac{1}{4}$ (b) $\frac{2}{5}$ (c) $\frac{7}{10}$

8 Convert the following to decimal fractions giving answers correct to 2 decimal places:

  (a) $\frac{1}{7}$ (b) $\frac{1}{9}$ (c) $\frac{2}{3}$

9 Express the following as percentages correct to 1 decimal place.

  (a) $\frac{5}{8}$ (b) $\frac{5}{9}$ (c) $\frac{3}{16}$

10 Express the following as fractions in lowest terms:

  (a) $8\frac{1}{3}\%$ (b) 6%   (c) 15%

11 Calculate the following:

  (a) $\sqrt{6400}$ (b) $\sqrt{0.0121}$

12 Which of the following is nearest in value to $\sqrt{4113}$?

  A. 20   B. 60   C. 200   D. 600   E. 2000

13 Which of the following is nearest in value to $\sqrt{0.0076}$?

  A. 0.003   B. 0.008   C. 0.03   D. 0.08   E. 0.3

14 Which of the following is nearest in value to $\sqrt[3]{0.0091}$

  A. 0.2   B. 0.3   C. 0.02   D. 0.03   E. 0.002

15 Which of the following is nearest in value to $\sqrt[3]{67\,152}$?

  A. 4   B. 40   C. 400   D. 4000   E. 40 000

> Tell your teacher you have finished this unit

UNIT 2
Holidays and Foreign Exchange

A Reading tables

Costa Brava HOLIDAYS Lloret

7, 10, 11, & 14 Nights from £170

This table is taken from a brochure which gives details of package holidays abroad.

The table gives prices of holidays at a hotel in Spain. The prices include air travel from Glasgow Airport.

Example

Find the cost of a holiday from Thursday, 13 September to Thursday, 27 September. The **departure date** is between July 17 and September 17. 14 nights are spent at the hotel.
Cost is £230.

Direct Jet Flights to Carona or Barcelona from	GLASGOW (Abbotsinch)			
Hotel	CLIPPER			
Holiday Duration **Nights in Hotel**	8 days 7 nights	11 days 10 nights	12 days 11 nights	15 days 14 nights
Departure Day Departure Time	Thur 15.50	Thur 15.50	Sun 09.10	Thur 15.50
Return Day Return Time UK	Thur 14.50	Sun 16.00	Thur 14.50	Thur 14.50
Deps Between April 14 – April 30	£ 170	£ 185	£ 193	£ 205
May 1 – May 14 Oct 16 – Oct 30	175	190	198	210
May 15 – June 18 Oct 1 – Oct 15	180	198	205	215
June 19 – July 1	188	205	213	223
July 17 – Sept 17	193	210	220	230
Sept 18 – Sept 30	188	205	213	225
FLIGHT CODE	A27	A27	A23	A27

Exercise

1 The table opposite shows prices of holidays in Majorca. The prices include air travel from Glasgow.

(a) Find the cost of a holiday from Monday, 21st May to Monday, 28th May.
(b) Find the cost of a holiday from Monday, 25th June to Friday, 6th July.
(c) Find the cost of a holiday for *two* adults from Monday, May 14th to Monday, May 21st.

Direct Jet Flights to Palma Airport from	GLASGOW (Abbotsinch)		
HOTEL	JAVA		
Holiday Duration **Nights in hotel**	7 days 7 nights	11 days 11 nights	14 days 14 nights
Departure Day Departure Time	Sun/Mon 01.30	Mon 23.50	Sun/M 01.30
Return Day Return Time UK	Mon 08.50	Fri 15.05	Mon 08.50
Deps Between April 14 – April 30	£ 170	£ 188	£ 185
May 1 – May 14 Oct 16 – Oct 30	175	180	218
May 15 – June 18 Oct 1 – Oct 15	180	200	228
June 19 – July 16	188	208	235
July 17 – Sept 17	195	215	245
Sept 18 – Sept 30	188	208	238
FLIGHT CODE	A13	A14	A13

MADRID

BENIDORM

Prices in £'s	Hotel Luna, Benidorm								
Jet direct from	LUTON				GATWICK		MANCH	GLASGOW	
Nights in hotel	7	10	11	14	7	14	14	7	14
Take-off time	Sun 09.30	Mon 22.30	Thur 15.30	Sun 09.30	Sun 15.00	Sun 15.00	Sun 16.30	Mon 01.00	Mon 01.00
Home landing	Sun 15.00	Thur 21.00	Mon 04.00	Sun 15.00	Sun 20.00	Sun 20.00	Mon 08.00	Sun 23.30	Sun 23.30
23 April – 30 April	173	170	180	200	175	203	210	170	205
1 April – 22 April 1 May – 17 May 1 Oct – 31 Oct	178	180	185	208	180	210	218	175	213
29 May – 15 June 17 Sept – 30 Sept	185	188	193	215	188	218	225	182	220
18 May – 28 May	188	190	195	220	190	223	230	185	225
16 June – 5 July 10 Sept – 16 Sept	190	193	198	225	193	228	235	187	230
6 July – 9 Sept	193	198	203	230	195	233	240	190	235
First departure	22 Apr	23 Apr	19 Apr	22 Apr	22 Apr	22 Apr	22 Apr	23 Apr	23 Apr
Last departure	21 Oct	15 Oct	18 Oct	14 Oct	21 Oct	14 Oct	14 Oct	22 Oct	15 Oct

2 The table above shows the prices of holidays from four airports – Luton, Gatwick, Manchester, and Glasgow.

(a) When is the most expensive time to take a holiday?

(b) If someone goes on holiday during this time and spends 7 nights at this hotel, what will it cost him if he flies from Gatwick?

(c) For a single room, there is an extra charge of £1.75 per night. How much extra does a person pay for a stay of 7 nights?

(d) Look at the prices of holidays from Luton and Glasgow which last 7 nights.

These columns are taken from the table above.

How much cheaper is each holiday from Glasgow than the same holiday from Luton?

(e) Now look at the prices of holidays from Luton and Glasgow which last 14 nights. Which holidays are cheaper in this case?

By how much are they cheaper?

LUTON	GLASGOW
173	170
178	175
185	182
188	185
190	187
193	190
22 April	22 April
21 Oct	21 Oct

Continue with Section B

B Holiday comparison

Cheap winter holidays

Package tour operators have now cut the price of winter holidays abroad to very low levels indeed.

Prices are lowest when people are least likely to want to go on holiday, (for example the week immediately before Christmas).

Let us compare the 'cost' of sunshine in December with its 'cost' in July.

Exercise

1 Copy and complete the following:

December		July	
Cost of 4–day holiday at Hotel Carmen	= £138	Cost of 7–day holiday at Hotel Carmen	= £280
Cost per day	= £138÷4	Cost per day	= ▨ ÷ ▨
	= ▨		= ▨
Average daily sunshine	= 5 hours	Average daily sunshine	= 8 hours
Cost of an hour's sunshine	= £34.50÷5	Cost of an hour's sunshine	= ▨ ÷ ▨
	= ▨		= ▨

Which is cheaper – winter holiday sunshine or summer holiday sunshine?

Continue with Section C

C   Winter sports holidays

WINTER SPORTS

Exercise

1 Calculate the cost of taking a school party to Innsbruck for a Winter Sports holiday leaving London on the 20th December. The party is made up of the following persons:

Age group	Number of persons
Under 12 years	2
12–13 years	4
14 years	12
15 years	3
16–20 years	2
Paying adults	2

The table below shows prices for **15 year olds.** Prices include fares from London to the resort in Austria. One adult is allowed to travel free with every ten members of the party.

Prices for **15 year olds**	9 Days Rail/Air				
Centre	20 Dec. 11 Jan.	27 Dec. 4 Jan.	9 Feb. 17 Feb.	4 April 11 April	19 April
Achenkirch	£172	£176	£174	£178	£177
Innsbruck	£171	£174	£172	£174	£173
Kirchdorf bei St. Johann	£180	£183	£182	£184	£181
Neustift Kampl	£173	£175	£174	£179	£177
Pfunds Stuben	£174	£178	£177	£177	£177
	Under 12 years	less £18			
Adjustments	12–13 years	less £14			
for	14 years	less £12			
other age groups	16–20 years	plus £12			
	Paying adults	plus £18			

Copy and complete the following table:

Age group	Cost per person	Number of persons	Total cost (£)
Under 12 years 12–13 years 14 years	£171 – £18 = £153	2	306
15 years 16–20 years Paying adults	£171	3	513
		Total cost of holiday =	

2 Calculate the cost of taking the following party to Davos Schatzalp leaving on 9th February.

Age group	Number of persons
12–13 years	10
14–15 years	8
16–20 years	4
Paying Adults	1

The table below shows prices for **14–15 year olds**.
Prices include fares from London to the resort in Switzerland.

Prices for 14–15 year olds	9 Days Rail/Air				
Centre	20 Dec. 11 Jan.	27 Dec. 4 Jan.	9 Feb. 17 Feb.	4 April 11 April	19 April
Rochers de Naye	**£173**	**£175**	**£173**	**£176**	**£174**
Leysin	**£172**	**£176**	**£174**	**£180**	**£179**
Davos Schatzalp	**£179**	**£182**	**£181**	**£183**	**£180**
Disentis	**£178**	**£180**	**£178**	**£186**	**£185**
Adjustments for other age groups	Under 12 years 12–13 years 16–20 years Paying adults	less £16 less £14 plus £13 plus £18			

Copy and complete the following table:

Age group	Cost per person	Number of persons	Total cost (£)
12–13 years 14–15 years 16–20 years Paying adults	£181 – £14 = £167 £181	10	
		Total cost of holiday =	

Continue with Section D

D Skiing holidays

Skiing in Scotland

There are also well established skiing centres in Scotland.

Here is a table of charges at Glenski Centre.

	Under 18 years	Adults
All–inclusive cost for 6 days	£183.50	£220.50
Accommodation only for 6 days	£120.50	£157.50
Ski–school (instruction)	£28 for 6 days or £5 per day	
Ski–hire (skis, boots, sticks)	£24 for 6 days or £4.50 per day	
Chair lift and tows	£4 per day	

1 Calculate the cost for 1 person (under 18 years) for accommodation, ski–school, ski–hire, and use of the chair lift for 6 days.

By how much is this greater than the all–inclusive cost?

2 Calculate the cost for an adult for accommodation, ski–school, ski–hire, and use of the chair lift for 6 days.

By how much is this greater than the all–inclusive cost?

3 Find the cost of a holiday lasting 6 days for 11 people (2 of whom are adults) at the all–inclusive charge.

4 Five people (all under 18 years of age) have a 6–day holiday at Glenski. They have skiing instruction on 3 days only. Find the following total costs for the five people:

(a) accommodation,
(b) 3 days at the ski–school,
(c) 3 days ski–hire,
(d) 3 days use of the chair lift.

What is the total cost?

Continue with Section E

E Ready reckoners

Many people go abroad for their holidays. One of the most popular countries for British holiday–makers is Spain.

Before a British holiday–maker goes shopping in Spain he should exchange some of his pound notes for **pesetas** — the currency of Spain.

Country	Unit of currency		Number per £1
U.S.A.	dollar	($)	1.25
Austria	schilling	(Sch)	25
Belgium	franc	(Fr)	72
Denmark	krone	(Kr)	12.86
France	franc	(Fr)	10.95
Germany	deutschmark	(Dm)	3.57
Italy	lira	(Lira)	2215
Spain	pesetas	(Pta)	197
Sweden	krone	(Kr)	10.12
Switzerland	franc	(Fr)	3.02

You'll find this kind of information in the financial section of a newspaper. A travel agency would also give you this kind of information.

From the table we see that we obtain 197 pesetas for £1.
197 pesetas to the £ is called the **rate of exchange**.

The rates of exchange vary from time to time. Check with the financial section of a newspaper for an up–to–date list.

Here is a ready reckoner to help you change pounds to pesetas. The amounts in pesetas are rounded to the nearest peseta.

£	0.10	0.20	0.30	0.40	0.50	0.60	0.70	0.80	0.90
Pesetas	20	39	59	79	99	118	138	158	177

£	1	2	3	4	5	6	7	8	9
Pesetas	197	394	591	788	985	1182	1379	1576	1773

£	10	20	30	40	50	60	70	80	90
Pesetas	1970	3940	5910	7880	9850	11820	13790	15760	17730

Example Change £89 to pesetas.

$$
\begin{array}{ll}
\text{pounds} & \text{pesetas} \\
80 \longrightarrow & 15\,760 \\
\underline{9} \longrightarrow & \underline{1\,773} \\
89 \longrightarrow & 17\,533
\end{array}
$$

Exercise

Use the ready reckoner above to change the amounts below to pesetas.

1 £75   **2** £48.50   **3** £91.20   **4** £82.30

Here is a ready reckoner to help you change pesetas to pounds. The amounts are rounded to the nearest penny.

Pesetas	10	20	30	40	50	60	70	80	90
£	0.05	0.10	0.15	0.20	0.25	0.30	0.36	0.41	0.46

Pesetas	100	200	300	400	500	600	700	800	900
£	0.51	1.02	1.52	2.03 ·	2.54	3.05	3.55	4.06	4.57

Pesetas	1000	2000	3000	4000	5000	6000	7000	8000	9000
£	5.08	10.15	15.23	20.30	25.38	30.46	35.53	40.61	45.69

Exercise

Use the ready reckoner to find what the articles shown would cost in British money.

5

6 *Chansons de Paris* 900 Ptas

7 **CIELO** COLOR 35 450 Ptas

8 2000 Ptas

9 2540 Ptas

3050 Ptas

Continue with Section F

F Making ready reckoners

Copy and complete this ready reckoner to change pounds into French francs. Complete the middle row first by adding on 10.95 each time, since £1 is 10.95 French francs.

10.95 ÷ 10

£	0.10	0.20	0.30	0.40	0.50	0.60	0.70	0.80	0.90
Francs	1.10	2.19	3.29						9.86

£	1	2	3	4	5	6	7	8	9
Francs	10.95	21.90	32.85						98.55

10.95 × 10

£	10	20	30	40	50	60	70	80	90
Francs	109.5	219	328.5						985.5

Example

Using the ready reckoner change £13.20 to francs.

pounds		francs
10	⟶	109.50
3	⟶	32.85
0.20	⟶	2.19
13.20	⟶	144.54

Exercise

Using the ready reckoner change the amounts below to francs. Set out working as above.

1 £27   **2** £46.70   **3** £83.20   **4** £98.50

Copy and complete this ready reckoner to change French francs into pounds. Complete the bottom row by multiplying the middle row by 10. Then complete the top row by dividing the middle row by 10 and rounding to the nearest penny.

$0.91 \div 10 = 0.091$
≈ 0.09

Francs	1	2	3	4	5	6	7	8	9
£	0.09	0.18	0.27						0.82
Francs	10	20	30	40	50	60	70	80	90
£	0.91	1.82	2.73	3.64	4.55	5.46	6.37	7.28	8.19
Francs	100	200	300	400	500	600	700	800	900
£	9.10	18.20	27.30						81.9

$0.91 \times 10 = 9.10$

Exercise

Use the ready reckoner to find what the articles shown would cost in British money.

5
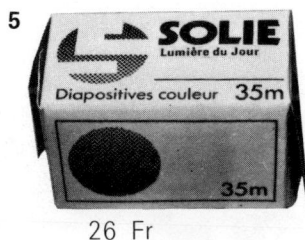
26 Fr

6

50 Fr

7
97 Fr

8

208 Fr

9
145 Fr

Continue with Section G

G Currency conversion

Pupils on a school cruise know that they will be calling at Copenhagen, Helsinki, Leningrad, and Stockholm.

It would be useful for them to have an exchange rate card to refer to when they go ashore at the ports.

The card would show exchange rates in Denmark, Finland, Russia, and Sweden and would help them convert the foreign prices into British money.

Here is a card which would do this.

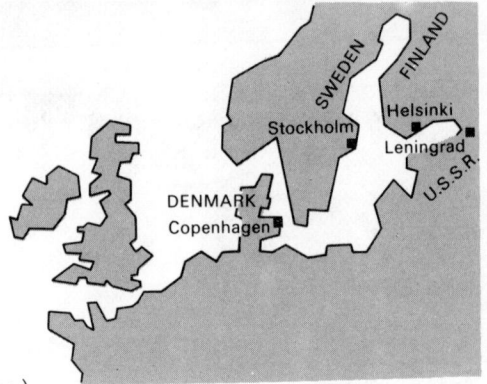

(Amounts are rounded to the nearest penny.)

EVERYMAN'S BANK					* * *
Amount of foreign currency	Denmark (Kroner) £1 = 12.86	Finland (Marrka) £1 = 6.41	U.S.S.R. (Roubles) £1 = 1.30	Sweden (Kroner) £1 = 10.12	Wherever you are going whether on business or holiday, we wish you a pleasant journey.
	£	£	£	£	
1	0.08	0.16	0.77	0.10	
2	0.16	0.31	1.54	0.20	
3	0.23	0.47	2.31	0.30	
4	0.31	0.62	3.08	0.40	
5	0.39	0.78	3.85	0.49	
10	0.78	1.56	7.69	0.99	
100	7.78	15.60	76.90	9.88	

To use the card we look at the first column and the column below the name of the country being visited.

Example

A transistor radio in Copenhagen costs 107 Kroner. How much is this in British money?

Danish Kroner		£
100	$\longrightarrow$	7.78
5	$\longrightarrow$	0.39
2	$\longrightarrow$	0.16
107	$\longrightarrow$	8.33

Exercise

1 A wooden carving in Helsinki costs 12 markka. How much is this in British money?

2 A souvenir teaspoon in Copenhagen costs 35 kroner. What is the equivalent amount in British currency?

3 A pupil exchanges £10 for roubles at the ship's bank before going ashore in Leningrad. If he spent 9 roubles how many did he have left? How much is this in British money?

4 Lemonade and cakes cost a pupil 15 kroner in Stockholm. If he also spent 35 kroner on a souvenir, how much had he left out of 80 kroner? If this amount could be changed into British currency how much would he get?

Continue with Section H

H Exchange rates

Country	Unit of currency		Number per £1
U.S.A.	dollar	($)	1.25
Austria	schilling	(Sch)	25
Belgium	franc	(Fr)	72
Denmark	krone	(Kr)	12.86
France	franc	(Fr)	10.95
Germany	deutschmark	(Dm)	3.57
Italy	lira	(Lira)	2215
Spain	peseta	(Ptas)	197

You are going to use the rates of exchange given in the table above.

Example

A British holiday-maker changes £50 into French francs. How many francs does he receive?

$$
\begin{array}{ll}
\text{£} & \text{Francs} \\
1 \longrightarrow 8.10 \\
50 \longrightarrow 8.10 \times 50 \\
 = 405
\end{array}
$$

Exercise

Do the following, setting out the working as in the example.

1 How many dollars would be obtained for £30?

2 How many lire would be obtained for £35?

3 A Spanish holiday–maker in Britain buys a souvenir which costs £5. What is the equivalent amount in Spanish currency?

4 Change £15 into deutschmarks.

5 Find how many Austrian schillings are given in exchange for £6.

Continue with Section I

I Progress check

1 Using the appropriate table on page 21, find the cost of a holiday from Glasgow, staying at the Hotel Java, from Monday 7th May, to Monday 21st May.

2 Using the appropriate table on page 25, find the cost of taking this party to Leysin leaving on 27th December.

Age group	Number of persons
12–13 years	10
14–15 years	10
Paying adults	2

3 Using the ready reckoner on page 27, change £15 to pesetas.

4 Using a ready reckoner on page 28, find what 3650 pesetas is worth in British money.

5 Using the ready reckoner on page 30, find what 7 roubles is equivalent to in British money.

6 Using the table on page 31, change £24 into Deutschmarks.

Ask your teacher what to do next

J Exchange rate changes

Find from the columns of a newspaper an up–to–date list of rates of exchange. Compare this list with the one given in Section E. You will find that there are many differences. Rates of exchange vary from time to time.

Here is a table which compares rates of exchange at March 1973 with rates at March 1985. Copy it and enter the present exchange rates.

Exchange rates for £1	U.S.A. (dollars)	Spain (pesetas)	Italy (lire)	France (francs)	Canada (dollars)
1973	2.54	146	1515	11.45	2.52
1985	1.25	197	2215	10.95	1.48
This year					

In 1973 £1 was worth **146 pesetas** while in 1985 it was worth **197 pesetas.** We say that the spending power of the pound had increased in Spain in 1985.

In 1973 the pound was worth **2.54 dollars**, while in 1985 it was worth **1.25 dollars**. We say the spending power of the £1 had dropped in the U.S.A. in 1985.

Exercise

1 From the table, find countries where the spending power of the pound had
   (a) increased and (b) dropped, between 1973 and 1985.

2 From the table, find countries where the spending power of the pound has
   (a) increased and (b) dropped, since 1985.

3 A student went to America in 1973 with £70 pocket money.
   How much was that in dollars? How much *less* would he have got in 1985 (in dollars)?

4 A girl went to Spain in 1973 with £20 pocket money.
   How much was that in pesetas? If she returned to Spain in 1985 with £20 how much more would she get (in pesetas)?

A change in the rate of exchange means a change in what the pound can buy in the foreign country. If there has been a drop in the rate of exchange it means that shopping is dearer for the holiday–maker.

Continue with Section K

K Surcharges

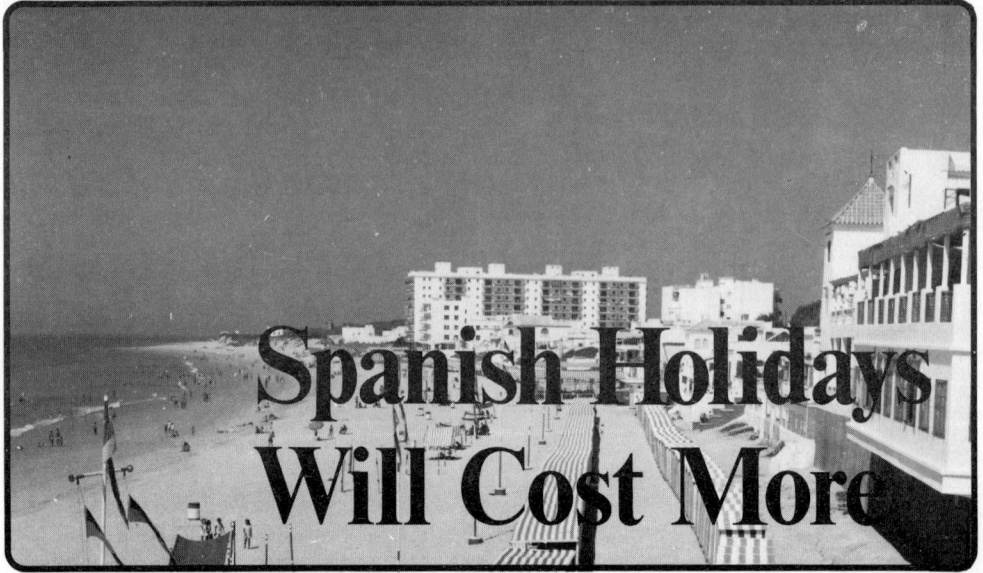

Headlines like these appeared when there was a sudden change in the number of pesetas that could be exchanged for £1.

People had booked their holidays abroad in advance so travel agents had to demand an extra payment to meet their costs.

The travel agents announced that there would be a 4% surcharge on all their holidays to Spain.

Example

In Section A we found that a 15-day holiday to Spain cost £192. How much would it cost with the 4% surcharge?

Original cost of holiday = £192

$\quad$ 1% of £192 = £1.92

$\quad$ 4% of £192 = 4 × £1.92

$\quad$ so surcharge = £7.68

New cost of holiday = £199.68

1% means $\frac{1}{100}$

So 1% of £192 = £192 ÷ 100

Exercise

1   Find the new cost of a holiday in Spain after the addition of a 4% surcharge on a holiday costing £178.

2   The cost of a holiday in Spain for a family of four was £836. Find the additional cost caused by a 4% surcharge.

3   Find the new cost of a holiday in Majorca after the addition of a 4% surcharge on a holiday costing £264.

Continue with Section L

L Conversion from foreign currency

A watch costs 170 francs in Switzerland. How much is this in British money, to the nearest penny, if the rate of exchange is 3.02 francs to the £?

Swiss francs £
3.02 ⟶ 1

1 ⟶ $\dfrac{1}{3.02}$

170 ⟶ $\dfrac{170}{3.02}$

$= 56.29$

Cost in British money is £56.29
(to the nearest penny).

Exercise

Find the cost in British money of each of the following, to the nearest penny. Set out your answers as in the example.

1 A musical box costs 34 Swiss francs.   (3.02 → £1)

2 A long-playing record costs 35 French francs.   (10.95 Fr → £1)

3 A camera costs 88.50 Dm in Germany.   (3.57 Dm → £1)

4 A scarf costs 27 kroner in Sweden.   (10.12 Kr → £1)

5 A meal costs 615 Fr in Belgium.   (72 Fr → £1)

6 A cinema ticket costs 58 Sch in Austria.   (25 Sch → £1)

7 A bullfight ticket costs 285 pesetas in Spain.   (197 Ptas → £1)

8 A pair of gloves cost 8100 lire in Italy.   (2215 lire → £1)

9 A bus tour of Rome costs 8000 lire.

Continue with Section M

M   Conversion table

Copy and complete this ready reckoner for changing Austrian schillings to pounds.
(28 Sch → £1) Round the amounts to the nearest penny.

Hint   Complete the bottom row first, and start by converting 90 schillings to pounds.

Schillings	1	2	3	4	5	6	7	8	9
£									
Schillings	10	20	30	40	50	60	70	80	90
£									

Continue with Section N

N   Progress check

Exercise

1   Find the cost in British money of each of the following:

    (a) A cow bell costs 4.50 Swiss francs.   (3.02 Fr → £1)
    (b) Sunglasses cost 35 Fr in France.   (10.95 Fr → £1)
    (c) A leather spectacle case costs 3200 lire in Italy.   (2215 lire → £1)

2   While a lady is on holiday in Spain the rate of exchange drops from 195 pesetas to the £ to
170 pesetas to the £. How much less will she get (in pesetas) if she now changes £50 to
pesetas?

3   Find the new cost of a £175 holiday when there is a surcharge of 6%.

Tell your teacher you have finished this unit

UNIT 3
Rounding and Scientific Notation

A Range to the nearest ten

Girls escape as car falls 160 m

It is unlikely that the girls fell exactly 160 metres. The reporter who covered the accident is unlikely to have climbed down into the gorge with a tape measure.

The newspaper has given an **approximation** to the actual number. We would expect the actual number to be nearer 160 than 150 or 170.

On a number line the actual number is in this range:

If the number 160 is correct to the nearest ten

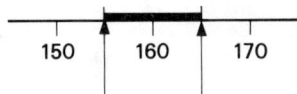

```
150     160     170
```

Then the actual number lies between **155** and **165**.

Example

The number 540 is correct to the nearest ten. In what range does the actual number lie?

Draw a number line going up in tens.
Mark the half-way points on either side of 540 and draw a line between them.

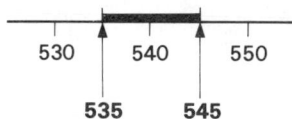

```
530     540     550
    535     545
```

The range is between 535 and 545.

Exercise

For each question draw a number line and find the range in which the actual number lies.

1   180 to the nearest ten.

2   910 to the nearest ten.

3   70 to the nearest ten.

4   600 to the nearest ten.

Continue with Section B

B   Other ranges

7000 protest over Arbroath Infirmary plan

It is unlikely there were exactly 7000 protestors.

The newspaper has given an **approximation** to the actual number.
We would expect the actual number to be nearer 7000 than 6000 or 8000.
On the number line the actual number is in this range:

| | 6000 | 7000 | 8000 |

If the number 7000 is correct
to the nearest thousand, the actual number lies between **6500** and **7500**.

Example

The number 6000 is correct to the nearest hundred. In what range does the actual number lie?

Draw a number line going up in hundreds.

| | 5900 | 6000 | 6100 |

Mark the half-way points on either side of 6000 and draw a line between them.

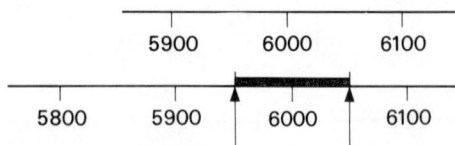

| | 5800 | 5900 | 6000 | 6100 |

The range is between 5950 and 6050.

Exercise

For each question draw a number line and find the range in which the actual number lies.

1   12 000 to the nearest thousand.
2   7800 to the nearest hundred.
3   63 000 000 to the nearest million.
4   60 000 to the nearest ten thousand.
5   700 000 to the nearest hundred thousand.
6   2000 to the nearest hundred.

Continue with Section C

C Rounding whole numbers

A certain Examination Board reports that the total number of candidates sitting Ordinary Grade Engineering Drawing was **11 907.**

This is a much more precise figure than is required for most purposes.

We could say that there were about 12 000 candidates
OR, if we want to be more exact, we could say there were about 11 900.

12 000 is the number of candidates **correct to the nearest thousand.**
11 900 is the number of candidates **correct to the nearest hundred.**
11 910 is the number of candidates **correct to the nearest ten.**

Example

Express 12 742 correct to the nearest hundred.

Draw a number line going up in hundreds.
Mark the half-way points.

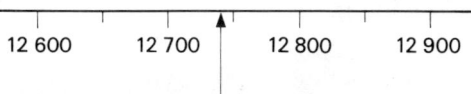

| 12 600 | 12 700 | 12 800 | 12 900 |

Roughly arrow 12742 and read off the nearest hundred, which is 12 700.

You may sometimes find that the number you are given comes exactly on one of the half–way points.

Example

Express 8450 correct to the nearest hundred.

When this happens you could choose either 8400 or 8500.
In this Unit we will choose the higher one, that is 8500.
So 8450 correct to the nearest hundred is 8500.

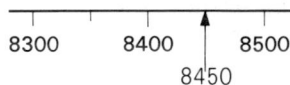

| 8300 | 8400 | 8500 |

8450

Exercise

For each question draw a number line.

1 Express 8375 correct to the nearest hundred.

2 Express 16 420 correct to the nearest thousand.

3 Express 748 correct to the nearest ten.

4 Express 157 432 correct to the nearest ten thousand.

5 Express 750 correct to the nearest hundred.

6 Express 7498 correct to the nearest thousand.

7 Express 7501 correct to the nearest thousand.

8 Express 12 345 678 correct to the nearest million.

Continue with Section D

D Rounding answers

In this section you are going to work out answers to problems and give your answers to a stated degree of accuracy.

Example

How many seconds (to the nearest million) are there in a year of 365 days?

Number of seconds in 1 minute = 60
Number of seconds in 1 hour   = $60 \times 60 = 3600$
Number of seconds in 1 day   = $3600 \times 24 = 86\,400$
Number of seconds in 1 year   = $86\,400 \times 365$
                              = $31\,536\,000$

```
|_____|_____|
31 000 000                    32 000 000
```

We could say:
The number of seconds in a year
is 32 million (to the nearest million).

Exercise

1 A box of pins contains 240 pins. Calculate, to the nearest thousand, the number of pins in 60 boxes.

2 A box of sweets contains 28 sweets. How many sweets, to the nearest thousand, will be needed to fill 150 boxes?

3 You eat three meals each day. How many meals, to the nearest hundred, do you eat in a year?

4 It is estimated that there are about 370 words on each page of a book of 84 pages. Calculate the number of words in the book giving your answer to the nearest thousand.

5 A packet of breakfast cereal weighs 270 grams. Calculate to the nearest thousand grams the weight of a carton containing 50 packets of cereal.

6 The actual number of people attending a Saturday First Division football match was 14 482. How would a newspaper record this attendance if it is given

(a) to the nearest ten, (b) to the nearest hundred, and (c) to the nearest thousand?

7 In one month a shop sells 283 television sets each of which cost £280. Calculate the total value of the sets, to the nearest hundred pounds.

8 A box of matches contains 49 matches. How many matches, correct to the nearest hundred, will be needed to fill 250 similar boxes?

Continue with Section E

E Rounding to the nearest whole number

An athletics coach records the time of an athlete doing a sprint back and forward across a gymnasium. His stopwatch records 16.9 seconds. He cannot be sure of his accuracy with the start and stop so he records the time **to the nearest whole number.**

16.9 to the nearest whole number is 17.

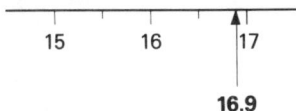

```
   15      16     |17
                  16.9
```

Example

Express 8.2 kilometres to the nearest kilometre.

8.2 kilometres to the nearest kilometre is 8 kilometres.

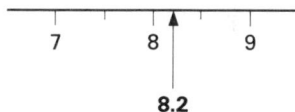

```
   7      8|      9
          8.2
```

Exercise

Express each of the following numbers to the nearest whole number.

1 11.6   **2** 19.9   **3** 141.2   **4** 60.5   **5** 4.8   **6** 86.4

Continue with Section F

F Decimal place notation

We now extend the methods to decimal fractions.

The pupils in the picture are working out the ratio of the **circumference** to the **diameter** of circles. We call this ratio π. What is the correct answer?

We could say that π is 3.1 or 3.14 or 3.142 or 3.1416 or 3.141 59 or 3.141 592. None of these are exact.

Saying $\pi = 3.1$ means that it is nearer to 3.1 than 3.0 or to 3.2.

Saying $\pi = 3.14$ means that it is nearer to 3.14 than to 3.13 or to 3.15.

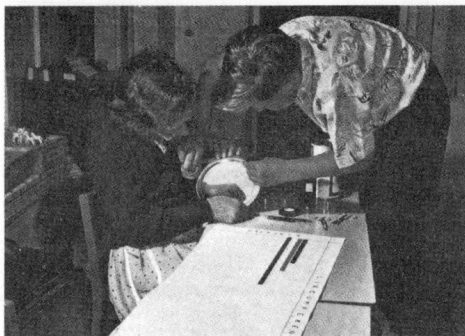

Do you remember what figures after the decimal point mean?

Units
tenths
hundredths
thousandths
ten thousandths
hundred thousandths
millionths

3. 1 4 1 5 9 2

We could say that 3.1 is correct to the nearest tenth.
We could say that 3.14 is correct to the nearest hundredth.
We could say that 3.141 592 is correct to the nearest millionth.

Instead we say that 3.**1** is correct to **1 decimal place.**
3.**14** is correct to **2 decimal places.**
3.**141 592** is correct to **6 decimal places.**
We use the contraction **dp** for **decimal places.**
So 3.**142** is correct to **3 dp.**

Example

The number 3.142 is correct to 3 dp. In what range does the actual number lie?

Draw a number line going up in hundredths.

Mark the half way points on either side of 3.142 and thicken the line between them.

The range is between 3.1415 and 3.1425.

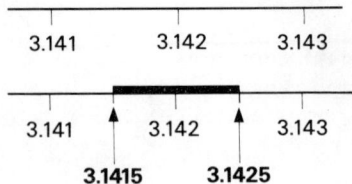

3.141   3.142   3.143

3.141   3.142   3.143
3.1415   3.1425

Exercise

For each question draw a number line and find the range in which the actual number lies.

1 41.6 correct to 1 dp.　　**2** 5.67 correct to 2 dp.　　**3** 7.12 correct to 2 dp.

4 8.90 correct to 2 dp.　　**5** 3.142 correct to 3 dp.　　**6** 7.526 correct to 3 dp.

Continue with Section G

G Rounding decimals

In Weir Academy there are 700 pupils of whom 100 are in years 5 and 6. If we work out the percentage of the pupils who are in 5 or 6 we get:

Fraction $= \frac{100}{700} = \frac{1}{7} = 0.142\,857\ldots = \frac{14.2857\ldots}{100}$

So the percentage is 14.2857 . . . .
This is much more exact than is usually required.

Example

Suppose we want the answer correct to 1 dp.
Draw a number line going up in tenths and mark the half-way points.

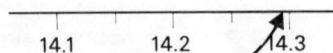

14.1   14.2   14.3

Roughly arrow 14.2857 . . . and read off the nearest tenth which is 14.3.

Example

Express the number 1.2365 correct to 3 dp.
Draw a number line going up in hundredths
and mark the half-way points.

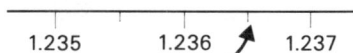

1.2365 is on a half-way point.
As before we choose the higher number.
So, 1.2365 correct to 3 dp is 1.237.

Exercise

For each question draw a number line.

1 Express 14.62 correct to 1 dp.

2 Express 2.436 correct to 1 dp.

3 Express 2.017 correct to 1 dp.

4 Express 8.173 correct to 2 dp.

5 Express 11.177 correct to 2 dp.

6 Express 7.8326 correct to 2 dp.

7 Express 2.1397 correct to 3 dp.

8 Express 1.9997 correct to 3 dp.

Continue with Section H

H Problems

Example

The stamp has length 4.1 cm and breadth
2.9 cm. What is its area correct to 1 decimal
place?

$$\text{Area} = L \times B$$
$$= 4.1 \times 2.9$$
$$= 11.89$$
$$\approx 11.9 \text{ cm}^2 \text{ correct to 1 dp}$$

$$\begin{array}{r} 41 \\ \times 29 \\ \hline 369 \\ 82 \\ \hline 1189 \end{array}$$

Exercise

1 Calculate the area of a desk top 0.6 m long and 0.3 m wide giving the answer correct to 1 dp.

2 Calculate the circumference of a circle of diameter, d, of 0.7 m giving your answer correct to 2 dp. (Circumference = πd; take $\pi = 3.14$)

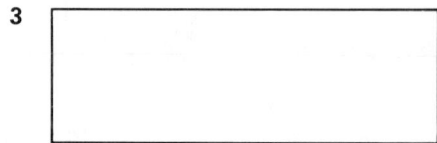

3

Measure in centimetres the length and breadth of the rectangle, each correct to 1 dp. Find its area in square centimetres correct to 1 dp.

4 Calculate the area correct to 1 dp of a semi–circle of radius, r, of 0.7 m. (Area=$\frac{1}{2}\pi r^2$; take $\pi=3.14$.)

5 In a school of 600 pupils, 100 are 16 years old or older. What percentage, correct to 1 dp, are 16 years old or older?

6 Express $\frac{2}{7}$ as a percentage correct to 1 dp.

7 Express $\frac{3}{11}$ as a percentage correct to 2 dp.

Continue with Section I

▋ Very large numbers

The mass of the earth is about
5 970 000 000 000 000 000 000 000 kg.

The number of seconds in a century is about 3 200 000 000.

Scientists and others have to work with large numbers like these. For example, a measure of length used in astronomy is called a parsec. It is equivalent to 30 700 000 000 000 kilometres approximately.

To be able to work with large numbers like this a shorthand method of writing them is used.

Look at this table:

$$100 = 10 \times 10 \qquad\qquad\quad = 10^2 \text{ (read as ten to the power 2)}$$
$$1000 = 10 \times 10 \times 10 \qquad\quad = 10^3 \text{ (read as ten to the power 3)}$$
$$10\,000 = 10 \times 10 \times 10 \times 10 \qquad = 10^4$$
$$100\,000 = 10 \times 10 \times 10 \times 10 \times 10 = 10^5$$

and so on.

Exercise

1 Copy the table above and complete the next three lines of the table for 1 000 000, 10 000 000, and 100 000 000.

Notice that the power of 10 is the same as the number of zeros after the 1.

Example

$$10\,000 = 10^4$$
4 zeros

$$10\,000\,000 = 10^7$$
7 zeros

Exercise

Write each of the following numbers as a power of 10.

2 1000

3 1 000 000

4 1 000 000 000

5 100 000 000

6 100 000 000 000 000

We can now put large numbers in shorthand by writing the number as a product.

Example

$5400 = 54 \times 100$
$= 5.4 \times 10 \times 100$
$= 5.4 \times 1000$
$\downarrow$
$= 5.4 \times 10^3$

Exercise

Write each of the following numbers in the way shown above.

7	8500	**8**	62 000 000	**9**	430 000
10	5 400 000 000	**11**	87 000 000 000 000	**12**	320 000 000

Notice that the first number in each answer lies between 1 and 10. Each answer is said to be in the form $a \times 10^n$ where

a is a number between 1 and 10

and n is 1, 2, 3, 4, ...

This way of writing numbers is called **standard form** or **scientific notation.**

Example

Write 275 000 000 in the form $a \times 10^n$ with $1 < a < 10$.

$275\,000\,000 = 275 \times 1\,000\,000$
$= 2.75 \times 100 \times 1\,000\,000$
$= 2.75 \times 100\,000\,000$

$= 2.75 \times 10^8$

To see a quick way of getting the answer study the diagram below.
We put the point here
in the $a \times 10^n$ form.

$2\,75\,000\,000 = 2.75 \times 10^8$

8 digits

Check this on the example at the top of the page.

Exercise

Write each of the following numbers in the form $a \times 10^n$, with $1 < a < 10$.

13	123 000	**14**	42 000 000	**15**	675 000 000
16	87 600	**17**	508 000	**18**	9460
19	17 600 000 000	**20**	1 134 000		
21	7 320 000 000 000 000 000 000 000 000 000				

Continue with Section J

J   Very small numbers

When dealing with very small quantities a microscope is often used.

The thickness of a piece of tissue paper is 0.000 03 m.

The size of a molecule of water is roughly 0.000 000 1 mm.

We can write very small numbers in the same $a \times 10^n$ way as we did for large numbers.

Look at this sequence of numbers

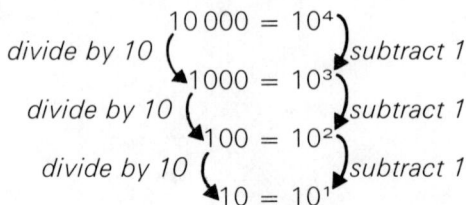

$$
\begin{array}{ll}
10\,000 = 10^4 \\
\text{\textit{divide by 10}} \Big(\qquad\qquad \Big) \text{\textit{subtract 1}} \\
1000 = 10^3 \\
\text{\textit{divide by 10}} \Big(\qquad\qquad \Big) \text{\textit{subtract 1}} \\
100 = 10^2 \\
\text{\textit{divide by 10}} \Big(\qquad\qquad \Big) \text{\textit{subtract 1}} \\
10 = 10^1
\end{array}
$$

If we continue to do the same, dividing the number on the left by 10 and subtracting 1 from the power of 10 on the right, we will get the following:

		Read as:
$10\,000$	$= 10^4$	
1000	$= 10^3$	
100	$= 10^2$	
10	$= 10^1$	10 to the power 1
1	$= 10^0$	10 to the power zero
0.1	$= 10^{-1}$	10 to the power negative 1
0.01	$= 10^{-2}$	10 to the power negative 2
0.001	$= 10^{-3}$	
0.0001	$= 10^{-4}$	

Exercise

1 Copy the table above and continue it for 0.000 01, 0.000 001, and 0.000 000 1.

Notice that for numbers less than 1 there is again an easy way of finding the answer. Study this diagram carefully:

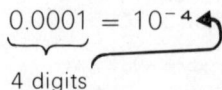

$$\underbrace{0.0001}_{\text{4 digits}} = 10^{-4}$$

We can now write small numbers in the form $a \times 10^n$ with $1 < a < 10$.

Example

$0.008 = 8 \times 0.001$

$\quad\quad = 8 \times 10^{-3}$

Example

$0.00072 = 72 \times 0.00001$

$\quad\quad\quad = 7.2 \times 10 \times 0.00001$

$\quad\quad\quad = 7.2 \times 0.0001$

$\quad\quad\quad = 7.2 \times 10^{-4}$

Exercise

Write each of the following numbers in the form $a \times 10^n$ with $1 < a < 10$.

2 0.0005   **3** 0.000097   **4** 0.0000084

Again there is a quick way of getting the answer.

Example

We look again at the second example above.

We put the point here
in the $a \times 10^n$ form

$0.00072 = 7.2 \times 10^{-4}$

Exercise

Write each of the following numbers in the form $a \times 10^n$ with $1 < a < 10$.

5 0.00062   **6** 0.000000752   **7** 0.000000000803

8 0.00524   **9** 0.000087   **10** 0.000000022

11 0.000512

Example

Write 0.00007236 in the form $a \times 10^n$ with $1 < a < 10$ and correct to 2 dp.

$\quad\quad\quad 0.00007236 = 7.236 \times 10^{-5}$

$\quad\quad\quad\quad\quad\quad\quad\quad = 7.24 \times 10^{-5}$ correct to 2 dp.

Exercise

Write each of the following in the form $a \times 10^n$ with $1 < a < 10$ and corrected as stated.

12 0.0001234   (2 dp)   **13** 0.0000678   (1 dp)

14 0.0052643   (2 dp)   **15** 0.000006175 (2 dp)

16 0.000000578941 (3 dp)   **17** 0.734618 (3 dp)

Continue with Section K

K Progress check

Exercise

1   The number of spectators at a football match is given as 15 600 correct to the nearest hundred. In what range does the actual number lie?

2   The density of helium in kilograms per cubic metre is 0.18 correct to 2 decimal places. In what range does the actual density lie?

3   Write 583 271 (a) to the nearest ten, (b) to the nearest hundred, and (c) to the nearest thousand.

4   Calculate the area of a rectangle 5.2 cm long and 2.7 cm broad, giving your answer to the nearest whole square centimetre.

5   Calculate the circumference of a circle of radius 2.4 cm giving your answer correct to 1 decimal place (circumference $= 2\pi r$; take $\pi = 3.14$).

6   Write 27.8435 correct to 2 dp.

7   Write 28.2465 correct to 3 dp.

8   Write each of the following numbers in standard form ($a \times 10^n$ notation with $1 < a < 10$)

(a) 52 000 000
(b) 7 200 000
(c) 5 000 000 000 000
(d) 0.000 063
(e) 0.000 005 2
(f) 0.008 73

9   Write each of the following in the form $a \times \mathbf{10^n}$ with $1 < a < 10$ with a correct to 2 dp in each case.

(a) 5 641 700
(b) 0.004 232
(c) 0.007 777

Ask your teacher what to do next

L. Calculations with positive indices

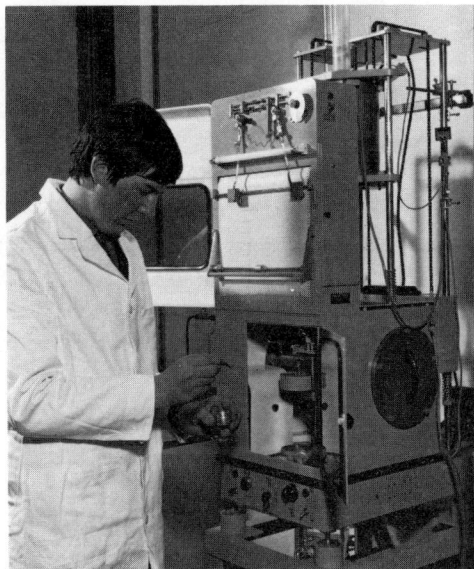

Scientists often need to multiply and divide large numbers. To do this we need to be able to deal with powers of 10.

Example

Can we write $10^3 \times 10^2$ as one power of 10?

$$10^3 \times 10^2 = (10 \times 10 \times 10) \times (10 \times 10) = 10^5$$

so
$$10^3 \times 10^2 = 10^5$$
$$\uparrow \qquad \uparrow \qquad \uparrow$$
$$3 + \quad 2 = \quad 5$$

To **multiply** powers of 10 together we **add** the **indices**.

Example

Can we write $10^5 \div 10^2$ as one power of 10?

$$10^5 \div 10^2 = \frac{10 \times 10 \times 10 \times 10 \times 10}{10 \times 10} = 10 \times 10 \times 10 = 10^3$$

So
$$10^5 \div 10^2 = 10^3$$
$$\uparrow \qquad \uparrow \qquad \uparrow$$
$$5 - \quad 2 = \quad 3$$

To **divide** powers of 10 we **subtract** the **indices**.

Exercise

Express each of the following as one power of 10.

1 $10^5 \times 10^3$ **2** $10^8 \times 10^2$ **3** $10^{12} \times 10^4$

4 $\dfrac{10^7}{10^2}$ **5** $\dfrac{10^{15}}{10^{12}}$ **6** $\dfrac{10^8}{10^5}$

7 $\dfrac{10^7}{10^4}$ **8** $\dfrac{10^9}{10^2}$ **9** $\dfrac{10^{12}}{10^4}$

Example

Express $(3 \times 10^2) \times (2 \times 10^3)$ in the form $a \times 10^n$ with $1 < a < 10$.

$$\begin{aligned}(3 \times 10^2) \times (2 \times 10^3) &= (3 \times 2) \times (10^2 \times 10^3) \\ &= 6 \times 10^{2+3} \\ &= 6 \times 10^5\end{aligned}$$

Exercise

Express each of the following in the form $a \times 10^n$ with $1 < a < 10$.

10 $(4 \times 10^2) \times (2 \times 10^2)$ **11** $(6 \times 10^3) \times (1.2 \times 10^4)$

12 $(4 \times 10^5) \times (2 \times 10^2)$ **13** $(1.3 \times 10^7) \times (5 \times 10^8)$

Example

Express $\dfrac{4.8 \times 10^7}{3 \times 10^4}$ in the form $a \times 10^n$ with $1 < a < 10$.

$$\begin{aligned}\frac{4.8 \times 10^7}{3 \times 10^4} &= \frac{4.8}{3} \times \frac{10^7}{10^4} \\ &= 1.6 \times 10^{7-4} \\ &= 1.6 \times 10^3\end{aligned}$$

Exercise

Express each of the following in the form $a \times 10^n$ with $1 < a < 10$.

14 $\dfrac{6 \times 10^6}{3 \times 10^3}$ **15** $\dfrac{8.4 \times 10^6}{4 \times 10^5}$

16 $\dfrac{4.8 \times 10^5}{4 \times 10^2}$ **17** $\dfrac{6.9 \times 10^{12}}{3 \times 10^5}$

18 $\dfrac{5.7 \times 10^5}{3 \times 10^2}$ **19** $\dfrac{(3 \times 10^4) \times (2 \times 10^3)}{5 \times 10^2}$

Continue with Section M

M Calculations with negative indices

When there are negative powers of 10 we proceed in the same way using the same rules.

Example

$$\begin{aligned}10^{-3} \times 10^{-5} &= 10^{-3+(-5)} \\ &= 10^{-8}\end{aligned}$$

5 to the left

Start here

Remember   To **multiply** powers of 10, **add** the indices.

Exercise

Express each of the following as a power of 10.

1 $10^4 \times 10^{-2}$ **2** $10^{-2} \times 10^{-3}$ **3** $10^{-7} \times 10^3$ **4** $10^{-4} \times 10^{-2}$

Example

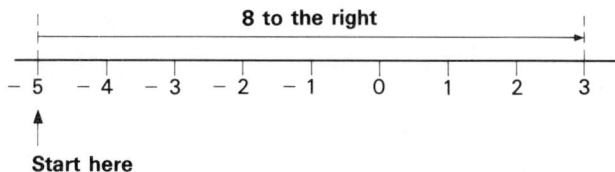

$$\frac{10^{-5}}{10^{-8}} = 10^{-5-(-8)}$$
$$= 10^{-5+8}$$
$$= 10^3$$

Example

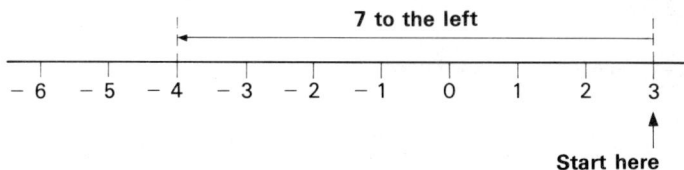

$$\frac{10^3}{10^7} = 10^{3-7}$$
$$= 10^{-4}$$

Remember To **divide** powers of 10, **subtract** the **indices**.

Exercise

Express each of the following as a power of 10.

5 $\dfrac{10^8}{10^{-3}}$ **6** $\dfrac{10^{-2}}{10^2}$ **7** $\dfrac{10^8}{10^{-9}}$ **8** $\dfrac{10^2}{10^{-10}}$ **9** $\dfrac{10^{-5}}{10^{-2}}$

Example

$$(3.1 \times 10^{-2}) \times (2 \times 10^{-6}) = (3.1 \times 2) \times (10^{-2} \times 10^{-6})$$
$$= 6.2 \times 10^{-2+(-6)}$$
$$= 6.2 \times 10^{-8}$$

Exercise

Express each of the following in the form $a \times 10^n$ with $1 < a < 10$.

10 $(2 \times 10^{-3}) \times (4 \times 10^{-7})$ **11** $(3 \times 10^{-7}) \times (2 \times 10^3)$

12 $(3 \times 10^{-2}) \times (3 \times 10^5)$ **13** $(4.1 \times 10^{-8}) \times (2 \times 10^{-9})$

14 $(3.1 \times 10^{-2}) \times (2.1 \times 10^{-9})$ **15** $(1.4 \times 10^{-8}) \times (2 \times 10^{12})$

Example

$$\frac{8 \times 10^{-3}}{2 \times 10^{-7}} = \frac{8}{2} \times \frac{10^{-3}}{10^{-7}}$$
$$= 4 \times 10^{-3-(-7)}$$
$$= 4 \times 10^{-3+7}$$
$$= 4 \times 10^4$$

Exercise

Express each of the following in the form $a \times 10^n$ with $1 < a < 10$.

16 $\dfrac{8 \times 10^{-3}}{2 \times 10^2}$ **17** $\dfrac{4 \times 10^2}{2 \times 10^{-3}}$ **18** $\dfrac{15 \times 10^{-9}}{5 \times 10^{-2}}$

19 $\dfrac{5.5 \times 10^{-9}}{5 \times 10^2}$ **20** $\dfrac{4.8 \times 10^5}{1.2 \times 10^{-3}}$ **21** $\dfrac{2.7 \times 10^{-8}}{1.35 \times 10^{-2}}$

Example

Express $(2 \times 10^{-3}) \times (8 \times 10^{-4})$ in the form $a \times 10^n$ with $1 < a < 10$.

$$\begin{aligned}
(2 \times 10^{-3}) \times (8 \times 10^{-4}) &= (2 \times 8) \times (10^{-3} \times 10^{-4}) \\
&= 16 \times 10^{-3+(-4)} \\
&= 16 \times 10^{-7} \\
&= 1.6 \times 10 \times 10^{-7} \quad [a \text{ must be between 1 and 10}] \\
&= 1.6 \times 10^{1+(-7)} \\
&= 1.6 \times 10^{-6}
\end{aligned}$$

Example

Express $\dfrac{1.5 \times 10^{-8}}{5 \times 10^{-2}}$ in the form $a \times 10^n$ with $1 < a < 10$.

$$\begin{aligned}
\frac{1.5 \times 10^{-8}}{5 \times 10^{-2}} &= \frac{1.5}{5} \times \frac{10^{-8}}{10^{-2}} \\
&= 0.3 \times 10^{-8-(-2)} \\
&= 0.3 \times 10^{-8+2} \\
&= 0.3 \times 10^{-6} \\
&= 3 \times 10^{-1} \times 10^{-6} \quad [a \text{ must be between 1 and 10.}] \\
&= 3 \times 10^{-1+(-6)} \\
&= 3 \times 10^{-7}
\end{aligned}$$

Exercise

Express each of the following in the form $a \times 10^n$ with $1 < a < 10$.

22 $(3 \times 10^{-2}) \times (4 \times 10^{-6})$

23 $(5 \times 10^{-7}) \times (3 \times 10^{12})$

24 $(6 \times 10^{-7}) \times (3 \times 10^{-8})$

25 $(2.4 \times 10^{-2}) \times (6 \times 10^{-5})$

26 $(3.8 \times 10^7) \times (5 \times 10^{12})$

27 $(2.5 \times 10^{-7}) \times (4 \times 10^{-8})$

28 $\dfrac{5.6 \times 10^{-7}}{8 \times 10^{-3}}$

29 $\dfrac{4.2 \times 10^{-11}}{7 \times 10^{-6}}$

30 $\dfrac{3.5 \times 10^{-10}}{5 \times 10^{-8}}$

31 $\dfrac{1.6 \times 10^3}{4 \times 10^{-6}}$

32 $\dfrac{3.6 \times 10^{-7}}{4 \times 10^{-3}}$

33 $\dfrac{2.4 \times 10^{-5}}{6 \times 10^{-8}}$

Continue with Section N

N Progress check

In each of the following questions simplify the expression and express the answer in the form $a \times 10^n$ where $1 < a < 10$.

1 $(2 \times 10^4) \times (4 \times 10^3)$

2 $(2.3 \times 10^{-3}) \times (3 \times 10^{-4})$

3 $\dfrac{8.4 \times 10^6}{3 \times 10^3}$

4 $\dfrac{8.1 \times 10^2}{3 \times 10^6}$

5 $(4 \times 10^6) \times (3 \times 10^6)$

6 $(6.4 \times 10^8) \times (3 \times 10^{-3})$

7 $\dfrac{4 \times 10^6}{8 \times 10^3}$

8 $\dfrac{8.1 \times 10^{-3}}{9 \times 10^3}$

Tell your teacher you have finished this unit

UNIT 4 Spending Money

A | Money is more convenient

In Tunisia, which is a country in North Africa, people often do their shopping in a local market. As you can see from the picture, the traders spread their goods across the street.

A system of **bartering** often operates. That is, the customer will swap goods he has made in exchange for the goods he wishes to buy.

A market place in Tunisia.

In Britain in former centuries a farmer might have paid for his new plough with 3 sheep, or the weaver might have paid for his groceries with a length of cloth.

Can you imagine your grocer's reaction if a farmer offered to pay his grocery bill with 1 cow?

It's much easier to use money. We all buy and sell goods using money.

Why work? Most adults in this country have a job. They do this so that they can earn **money**. Money allows us to buy food and clothes, and all the things we like to spend our money on.

Example

Here is a radio costing £9.95. Notice that it can also be bought by making 6 monthly payments of £1.78.

Buying in this way is called **hire purchase**.

Cash price of radio = £9.95
6 payments of £1.78 = 6 × £1.78
= £10.68

Hire purchase cost = £10.68
Cash price = £ 9.95
Extra amount paid = £ 0.73

PHILIPS AM

Exercise

In each of the following questions calculate the total cost of buying by **hire purchase**, then calculate the extra amount paid by using hire purchase.

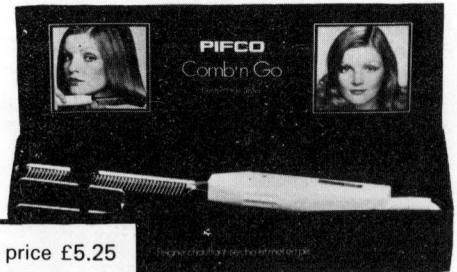

PIFCO Comb'n Go

1 Copy and complete:

Cash price of styler = £5.25
6 payments of £0.99 = 6 × £0.99
= £
Hire purchase cost = £
Cash price = £
Extra amount paid = £

Cash price £5.25
or
6 monthly
payments of 99 p

Do these in the same way.

2 Cash price £17.90
or
6 monthly payments of £3.32

3 Cash price £13.65
or
12 monthly payments of £1.31

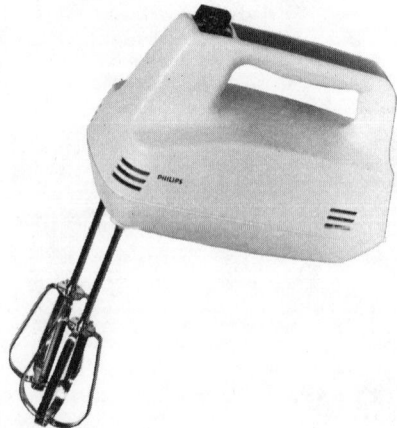

Continue with Section B

B Hire purchase: deposits

Very often the hire purchase customer is expected to pay a **deposit** and then monthly payments.

Example

A watch has a cash price of £24.50. It can also be bought on hire purchase by paying a deposit of 10% of the cash price and 6 monthly payments of £4.40. Calculate the total cost of buying the watch using hire purchase.

To solve this we have to find 10% of £24.50.

Remember that 10% means $\frac{10}{100} = \frac{1}{10}$.

To find $\frac{1}{10}$ of £24.50 we move the figures one place to the right.

£24.50
£ 2.45

Cash price of watch = £24.50
Deposit of 10% of cash price = £2.45
6 payments of £4.40 = 6 × £4.40
= £26.40
Total HP cost = £2.45 + £26.40
= £28.85

Note that **HP** is a commonly used abbreviation for **hire purchase**.

Exercise

Cash price £85.50 or 10% deposit plus 6 monthly payments of £15.30

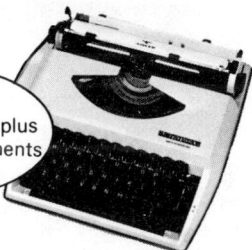

1 Copy and complete:

Cash price of typewriter = £▒
Deposit of 10% of cash price = £▒
6 payments of £15.30 = 6 × £▒
= £▒

Total HP cost = £▒ + £▒
= £▒

CASH PRICE
£94.50

DEPOSIT OF 10% of CASH PRICE +12 monthly payments of £8.20

2 Copy and complete:
Cash price of cycle = £▒
Deposit of 10% of cash price = £▒
12 payments of £8.20 = 12 × £8.20
= £▒

Total HP cost = £▒

VECTO FAN HEATER

CASH PRICE **£21.50**

10% DEPOSIT plus 6 monthly payments of £3.75

3 Find the total HP cost of the heater:

Continue with Section C

C Hire purchase and cash price

Example

A tent costs £299. It may also be bought using hire purchase by paying a deposit of 8% of the cash price and 6 monthly payments of £53. Calculate the total cost of buying the tent on HP, and show the cash cost and HP cost in a bar chart.

Cash price of tent	= £299
1% of £299	= £2.99
So 8% of £299	= 8 × £2.99
	= £23.92
6 payments of £53	= 6 × £53
	= £318
Total HP cost	= £23.92 + £318
	= £341.92

Notice that to find 8% of £299 we first find 1% of £299 by moving the figures two places to the right.
Then we multiply by 8.

£ 2 9 9 .
£   2 . 9 9

Cost (£)

[Bar chart with vertical axis marked 0, 80, 160, 240, 320, 400, showing two bars labelled "Cash price" and "H.P. price"]

Exercise

In each of the following calculate the total HP cost. Show the cash price and the HP price in a bar chart.

1 Cash price of record

player	= £120
1% of £120	= £
So 8% of £120	= 8 × £
	= £
12 payments of £10.80	= 12 × £10.80
	= £
Total HP cost	= £ + £
	= £

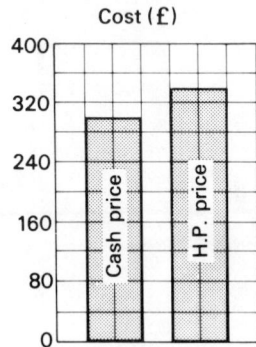

ZONY 20 MUSIC SYSTEM

£120

8% deposit plus 12 monthly payments of £10.80

2

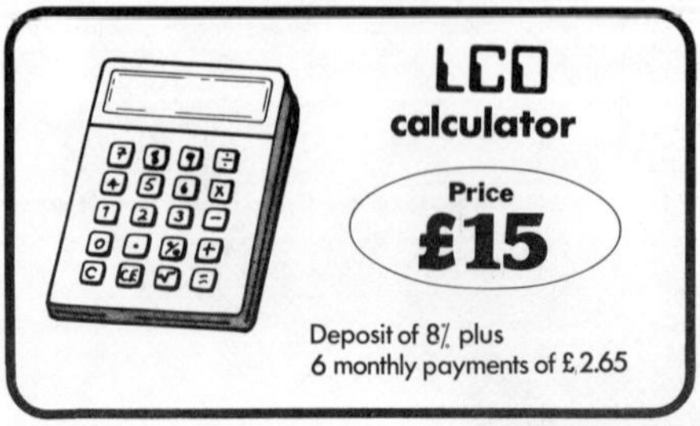

LCD calculator

Price **£15**

Deposit of 8% plus 6 monthly payments of £2.65

3

Outfit 36

ANKAMATIC 36

Price **£16.80**

10% DEPOSIT plus 6 monthly payments of £3.24

Continue with Section D

D Percentage increase in price

When we go to buy something we often find that the price has increased.

Here are some sports goods whose prices have increased. In each case we have to calculate the new cost (giving our answers to the nearest penny).

Example

A leather football which cost £15.40 has had its price increased by 6%. What is the new price?

Original price	= £15.40
1% of £15.40	= £0.154
So 6% of £15.40	= 6 × £0.154
	= £0.924
	≈ 92p (to the nearest penny)
New price	= £15.40 + 92p = £16.32

Exercise

1 A badminton racket which cost £10.80 has had its price increased by 4%. What is the new price?

Copy and complete:

Original price	= £10.80
1% of £10.80	=
So 4% of £10.80	=
	=
	≈
New price	=

2 A ten day holiday in Majorca which cost £250 has had its price increased by 8%. What is the new price?

3 A beach robe which cost £12.80 has had its price increased by 5%. Calculate the new price.

4 The prices of all the goods listed have to be increased by 12%. Calculate the new prices giving your answers correct to the nearest penny.

Safeway Sparkling MINERALS	**32**p
Corn Niblets or MEXICORN	**36**p
100g NESCAFE Granules	**£1.40**p
Kleenex Double TOILET ROLLS	**56**p
½ kg Kraft Soft MARGARINE	**45**p

Copy and complete:

Safeway minerals

Original price	= 32p
1% of 32p	= 0.32p
So 12% of 32p	= 12 × 0.32p
	= 3.84p
	≈ 4p (to the nearest penny)
New price	= ▉ p

Now do the others in the same way.

Continue with Section E

E Percentage increase

Exercise

1   A cable between two pylons is 120 m long. If it stretches by 3% of its length, what is its new length?

2   A football team gets an average crowd of 24 000 for home games. If they have a successful run in which they win eight games in a row and their support increases by 15%, what is the new average crowd for their home games?

3   A man weighs 86 kg. However he eats too much and after 1 year his weight has gone up by 12%. What is his new weight? (Answer to nearest kg.)

4   British Leyland have to increase the price of their Austin Montego by 5%. What is the new price if the original price was £6480?

5   A school pupil runs a paper round with a weekly turnover of £45 (that is, his total sales are £45). He has a sales campaign and increases his turnover by 8%. What is his new turnover?

If he makes a profit of 12% of his new turnover how much money does he now make? (Answer to the nearest penny.)

Continue with Section F

F Percentage reduction in price

NEW YEAR SALE

You are a shop assistant. The shop has a sale. All prices are reduced by 8%. Take the advert for the local paper and change the prices ready for the sale. Here is what you would have to do.

Example

Ladies raincoats

Original cost = £27.50
1% of £27.50 = £0.275
So 8% of £27.50 = 8 × £0.275
= £2.20
Sale price = £27.50 − £2.20
= £25.30

LADIES FASHIONS

Tremendous Value in ...

LADIES' RAINCOATS
Terylene and Cotton mixture: ¾- or full-length, with warm quilted lining: choice of turquoise, beige, or navy; sizes 12-20. ALL ONE PRICE £27.50 £25.30

"RAINSEAL" ¾-LENGTH COAT
with Borge fabric trimmed with black double-breasted styling with ½ belt. Sizes 12-18 .... £25.50 £23.46

Exercise

1 Work out the reduced price of the 'Rainseal' coat (was £25.50) and check that it agrees with the above answer of £23.46.

2 Calculate the sale prices for all these menswear items if all goods are reduced by 6%. (Answer to nearest penny where necessary.)

MENSWEAR

"QUEST" 2-PIECE SUITS
3-button styling: sizes 36-44; pure new wool from £40. Terylene/Sarril at **£68**

CASUAL JACKETS
Modern double-breasted styling in imitation suede; sizes small, medium, large **£42**

MEN'S RAINCOATS
In Polyester/ Cotton; colours navy or burnt oak; sizes 36-40 .... **£37.40**

MEN'S SHIRTS
By Black Bear
Selection of up-to-the-minute styles in fashionable colours; sizes 14½-16½; from **£9.90**

YOUNG MEN'S FASHION KNITWEAR
In Easy-care Acrilan; Turtle neck style; choice of colours; sizes small and medium........... **£11.10**

"WOLSEY" MEN'S SINGLETS & BRIEFS
In white Cotton Vincel; sizes small, medium, large, and extra large .... **£1.80** each

Continue with Section G

G Discount

Many people when they go shopping don't pay over any cash when they buy goods. They simply say 'put it on my account'. The shopkeeper adds up all their purchases for a month and sends them an account.

Very often you will even get a discount – that is, some money taken off your bill. Look at the bill opposite.

1% of £62.50 = £0.625
So 5% of £62.50 = 5 × £0.625
= £3.125
≈ £3.12

CHARGE ACCOUNT

A Charge Account at Silverbergs is the modern family's way to buy. You can shop at any time, in any of our branches and all your purchases are payable approximately four months later and you still get 5% discount.

EXAMPLE:		
	Ladies Dress	£21.00
	Skirt	£15.00
	Blouse	£12.50
	Shoes	£14.00
		£62.50
	Less 5% discount	£ 3.12
		£59.38

Payable 4 months after purchase

Note *When a discount works out as £3.125 a store would normally only give £3.12 discount.*
Discount should be rounded **down** *to the nearest penny.*

Exercise

Copy and complete the following two accounts:

1
Trousers	£16.90
Sweater	£ 8.20
Dress	£20.14
Pyjama suit	£ 8.26
	£▓▓▓
Less 5% discount	£▓▓▓
	£▓▓▓

Working
1% of £▓▓▓ = £▓▓▓
So 5% of £▓▓▓ = 5 × £▓▓▓
= £▓▓▓
≈ £▓▓▓

2
Training shoes	£14.80
Pullover	£ 8.84
Safari jacket	£14.14
Briefs pack	£ 4.32
	£▓▓▓
Less 5% discount	£▓▓▓
	£▓▓▓

Example

Here is a mock-up of an invoice for an article costing £5.60 which is given a 5% discount. Note the terms **gross amount** (this is the initial cost) and **net amount** (this is the cost with discount taken off).

Date	Invoice No.	Store Code	Gross Amount	Discount	Net Amount
5.08.85	3855	212	5.60	0.28	5.32

Exercise

Below are mock–ups of three different monthly accounts. For each account calculate the discount and the net amount.

3

Date	Invoice No.	Store Code	Rate of Discount	Gross Amount	Discount	Net Amount
1.05.85	16413	117	5%	46.50		

4

Date	Invoice No.	Store Code	Rate of Discount	Gross Amount	Discount	Net Amount
9.07.85	58561	0	4%	13.10		

5

Date	Invoice No.	Store Code	Rate of Discount	Gross Amount	Discount	Net Amount
12.09.85	38552	202	8%	18.30		

Continue with Section H

H Percentage reduction

Exercise

1 A car depreciates in value – that is, as it gets older it is worth less money. A new car can depreciate by as much as 30% in its first year. If this Vauxhall Astra costs £6320 when new, how much is it worth 1 year later if it has depreciated by 30%?

Note   The sensible way to calculate 30% is to calculate 10% first, and then multiply by 3.

2 An iceberg loses 8% of its weight by melting in a year. How much will an iceberg weigh after 1 year if its original weight was 550 000 kg?

3 The speed of a certain ship is reduced by 12% when it is fully laden. If it can travel at 15 knots when unladen, what is its speed when laden?

4 I used to be able to type at 60 words per minute. I am now out of practice and my speed has dropped by 8%. What is my present speed (to the nearest number of words per minute)?

5 A man weighs 106 kg and is told by his doctor to lose weight. If he reduces his weight by 18%, what is his new weight? (Answer to the nearest kg.)

Continue with Section I

Percentage composition from pie charts

Example

The diagram show the percentages of boys and girls in a town. If the total number of boys and girls is 25 000 how many girls are there and how many boys?

Total number of boys and girls = 25 000
1 % of 25 000   =   250
So 55 % of 25 000   = 55 × 250
                               = 13 750
Number of girls   = 13 750
So number of boys   = 25 000 − 13 750
                               = 11 250

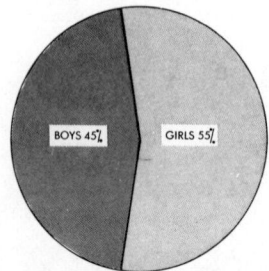

BOYS 45% GIRLS 55%

Note   *this way of showing information is called a **pie chart**.*

Exercise

1 Here is a pie chart showing petrol sales by grades in Scotland. If the total quantity sold was 1 180 000 tons calculate the quantity sold of each grade.

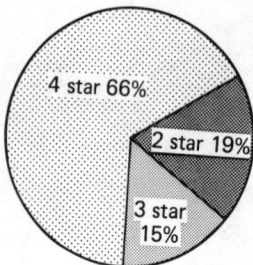

4 star 66%
2 star 19%
3 star 15%

Copy and complete:

Total quantity sold = 1 180 000 tons
1% of ▨▨▨▨▨   = ▨▨▨▨▨ tons
2 star
19% of 1 180 000 = 19 × ▨▨▨▨▨
                  = ▨▨▨▨▨ tons

Now calculate the sales for 4 and 3 star petrol.

2 In 1976 the Scottish Gas Board produced 348 million therms and sold them to domestic, commercial, and industrial consumers as shown in the diagram. Calculate how many million therms each group consumed (to the nearest million therms).

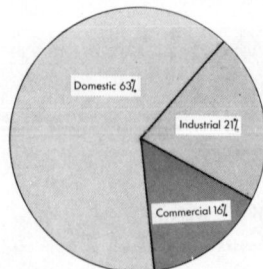

Domestic 63%
Industrial 21%
Commercial 16%

Continue with Section J

J Drawing pie charts

This unit is about *Spending Money*. You know how you spend your own money, but do you know how everyone else spends their money?
In a certain year in Britain we spent our money as shown below

Food, drink, tobacco	36%
Housing, fuel, light	27%
Clothing	12%
Household goods	6%
Cars	4%
Miscellaneous	15%

This can be shown in a pie chart. It is easier to show information on a pie chart if we use a circle which is divided into 100 parts as shown above.

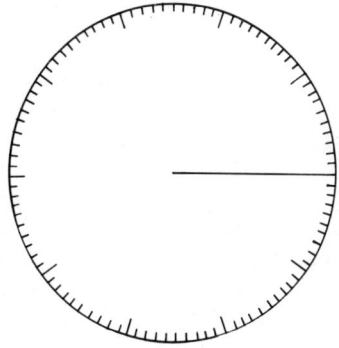

Exercise

For this exercise you need a pie chart sheet. Ask your teacher for one.

1 On the first chart count round 36 divisions like this — make the sector as shown.

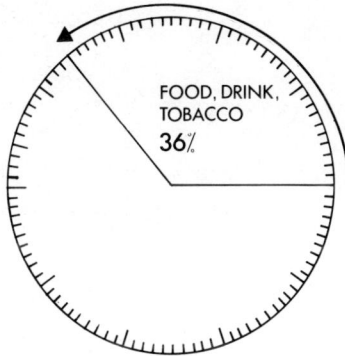

Add the next sector by counting round 27 divisions on the same chart like this.

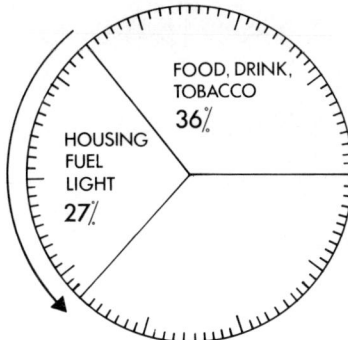

Now complete the remaining sectors in the chart.

2 Using the next chart draw a pie chart to show the following information:

Sources of energy for Britain in 1983

Source	Percentage
Oil	39
Coal	33
Gas	21
Nuclear/hydro	7

3 East Kilbride was the first and largest of Scotland's five new towns. When study- ing a new town it is important to know **where** the people have come from, since most of them have come from other towns.

This pie chart shows where the new tenants came from. You can see that most of East Kilbride new tenants came from Glasgow.

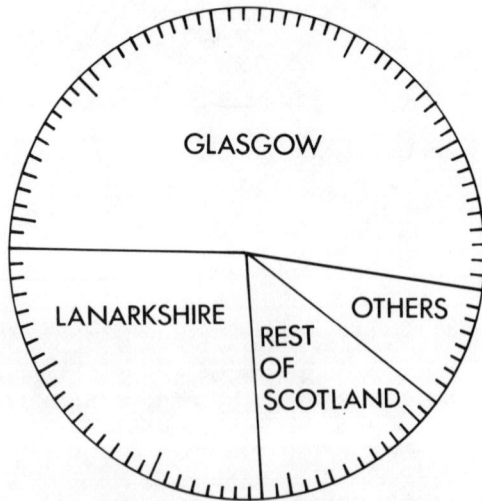

List the various areas with their percentages. Copy and complete:

Glasgow	52%
Lanarkshire	%
Rest of Scotland	%
Others	%

Continue with Section K

K Percentage calculation and drawing pie charts

Since you don't get paid for going to school your spending money will be either

pocket money, or
money you earn in part-time work.

Here are the results of a survey done in a school showing the amount of pocket money received by a sample of forty 14 to 15 year old pupils.

Pocket money (pence)	Tally	Frequency	Percentage
0–49	|||	3	7.5 ◄
50–99	₩₩ ₩₩	10	
100–149	₩₩ ₩₩ ₩₩ ||||	19	
150 and over	₩₩ |||	8	
Total		40	

To show this information in a pie chart we must change the frequencies into percentages.

Example

(0–49) group There are 3 in this group out of a total sample of 40.

$$\frac{3}{40} = 7.5\%$$

Working

$$\frac{3}{40} = 0.075$$

$$= \frac{7.5}{100}$$

$$= 7.5\%$$

Exercise

1   Copy and complete the table above. Show the information on a pie chart.

2   Carry out a survey of numbers of brothers and sisters.
List them as 0, 1, 2, 3, 4 or more and draw a pie chart.

   Note   *It will make your calculations easier if you choose a sample of 20 or 25 pupils.*

3   Think up a survey of your own and ask your teacher's permission to carry it out. Examples could be (a) colour of eyes, or (b) shoe sizes, or (c) favourite pop star. Draw a pie chart.

Continue with Section L

L Progress check

Exercise

1 A cassette recorder can be bought for a cash price of £37 or for a deposit of 12% of the cash price and 12 monthly payments of £3.33.

(a) What is the total HP cost?

(b) How much greater is this than the cash price?

2 A man earns £96 per week. If his weekly wage is increased by 20% what will his new wage be?

3 In a summer sale all prices are reduced by 15%. What will be the reduced price of a suit originally priced at £76?

4

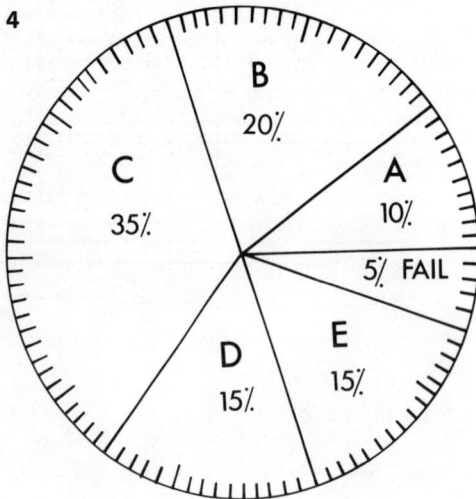

In a certain examination candidates are given grades A, B, C, D, E, or FAIL.

The pie chart shows the distribution of the 240 mathematics candidates presented by one school.

(a) How many candidates were given a C grade?

(b) How many candidates were given a D grade?

5 A jar of coffee costs £1.68. The price is increased by 12%. Calculate the new price (to the nearest penny).

Ask your teacher what to do next

M Calculation of hire purchase terms

You are the boss of a radio and television shop. You have a casette tape recorder with a *cash price of £49*. You have to calculate the hire purchase terms.

Here is what you would have to do.

(a) The HP cost is to be 15% above the cash price.

Cash price   = £49
1% of £49   = £0.49
So 15% of £49 = 15 × £0.49
               = £7.35

This is the *extra* your customers will have to pay if they buy the tape recorder on HP.

Total HP cost = £49 + £7.35
              = £56.35

(b) The deposit is to be 10% of the cash price.

Cash price   = £49
10% of £49   = £4.90
So deposit   = £4.90

(c) The balance is to be paid by 12 equal monthly amounts.

Total HP cost = £56.35
Deposit   = £4.90
Amount still to be paid = £56.35 − £4.90
                        = £51.45
Monthly payment = £51.45 ÷ 12
                = £4.29 (to the nearest penny)

$$\begin{array}{r} 4.287 \\ 12\overline{)51.450} \\ 48 \\ \hline 34 \\ 24 \\ \hline 105 \\ 96 \\ \hline 90 \end{array}$$

Notice that 4.287 is nearer to 4.29 than to 4.28.

Your terms are:

Deposit of £4.90, and 12 monthly payments of £4.29.

Exercise

1 A stereo system has a cash price of £70.

  (a) Selling it on HP you want to make an extra 15%. Calculate the total cost on HP.
  (b) The deposit is 10% of the cash price. What is the deposit?
  (c) The balance is spread equally over 12 months. How much must you charge for each payment?

Copy and complete:

(a) Cash price   = £70
     1% of £70   = £▓
     So 15% of £70 = £▓
     Total HP cost = £▓

(b) Cash price   = £▓
     So 10% of £70 = £▓
     Deposit   = £▓

(c) Total HP cost   = £▓
     Deposit   = £▓
     Amount still
     to be paid   = £▓
     Monthly payment = £▓

2 A portable television set has a cash price of £65.

  (a) On HP terms the dealer wishes to make an extra 20% on the cash price. What is the total cost of the TV set using HP?
  (b) The deposit is 10% of the cash price. How much is the deposit?
  (c) Calculate the cost per month for 10 equal monthly payments.

3 A Mini Metro has a cash price of £5380. The HP cost is 5% higher. Calculate the total HP cost.

A deposit of 10% of the cash price is charged. Calculate the deposit and the monthly repayments if they are spread over 24 equal payments.

4 A Fiat Panda has a cash price of £4350. The HP cost is 6% higher. Calculate the total HP cost.

A deposit of 12% of the cash price is charged and the amount still owing is divided into 24 equal amounts. State the HP terms.

Continue with Section N

N Commission

Before you start *spending* money you have to *earn* some money.
Here are adverts for 2 different jobs. Which one would you choose?

PROVISIONS CUTTER
COOPERS FINE FARE

£120 per week

Coopers Fine Fare require an experienced Provisions Cutter for their new Super Store in the Town Centre, East Kilbride.

We offer a starting salary of up to £120 per week, depending on degree of experience. Excellent working conditions and 40-hour week.

SALES REPRESENTATIVE

Energetic young man required to sell well known brand of foodstuffs.

Basic wage of £80 plus 1½% of weekly sales. Average weekly sales of existing sales representatives is £4800.

In the job of sales representative you are given a fixed amount *plus* an extra amount called a **commission**. The value of this commission depends on how hard you work and how many articles you sell. The advert says that average sales amount to £4800. We can calculate the wage as follows:

Fixed amount = £80
1½% of £4800 = £72 ◄
Total wage   = £152

1% of £4800 = £48
½% of £4800 = £24
1½% of £4800 = £72

Exercise

1 In the example above calculate the weekly wage if the sales were
   (a) £3200   (b) £6100

2 A salesman is engaged at a wage of £100 per week, together with a commission of 3% on all sales. Calculate his wage in a week in which he sold £3100 worth of goods.

3 A man is paid £3300, plus a commission of $2\frac{1}{2}$% on sales. Calculate his salary when sales are £74 800.

4 A man received an annual salary of £7200, together with a commission of 5% on all sales **over** £10 000 per annum. Calculate his annual salary in a year in which his sales amounted to £72 800.

   Note *In this case the commission is earned on £72 800–£10 000.*

Continue with Section O

O Increase or decrease as a percentage

Which of the two adverts is the better bargain?

Television set
£10 is saved out of £50.
This is a saving of $\frac{1}{5}$ or 20%.

Radio
£5 is saved out of £20.
This is a saving of $\frac{1}{4}$ or 25%.

The radio is the better bargain.

The best bargains are those with the *highest percentage saving*.

To work with percentages we must be able to change fractions like $\frac{9}{25}$, $\frac{4}{23}$, $\frac{5}{7}$, into percentages.

Example

$\frac{4}{23} \approx 0.173$

≈ 0.17 (to 2 decimal places)

$= \frac{17}{100}$

$= 17\%$

$$
\begin{array}{r}
\textit{Working} \\
0.173 \\
23\overline{)4.000} \\
\underline{2\,3} \\
1\,70 \\
\underline{1\,61} \\
90 \\
\underline{69} \\
21 \\
\end{array}
$$

We can do the working using:

1 a calculator,
2 long division (as shown on page 70)

Exercise

1 Write the following fractions as percentages (answer to the nearest whole number).

$$\frac{4}{5} \qquad \frac{3}{7} \qquad \frac{13}{19} \qquad \frac{21}{60} \qquad \frac{11}{75} \qquad \frac{2.5}{8.9}$$

Example

Recommended retail price (RRP) $= £180$
Special sale price $= £147$
Amount saved $= £33$

Fraction saved $= \dfrac{£33 \leftarrow \text{saving}}{£180 \leftarrow \text{original price}}$

≈ 0.183

≈ 0.18 (to 2 decimal places)

$= \dfrac{18}{100}$

$= 18\%$

ELECTRO FRIDGE

Model 45

OUR PRICE

£147

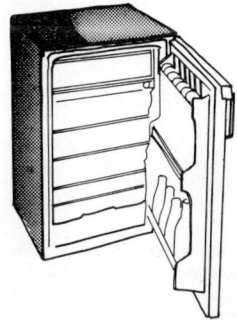

▶ 5.4 cubic feet ◀

Working

$$
\begin{array}{r}
0.183 \\
180)\overline{33.000} \\
180 \\
\hline
1500 \\
1440 \\
\hline
600
\end{array}
$$

So $\dfrac{33}{180} \approx 0.18 = \dfrac{18}{100}$

Recommended Retail Price (RRP) **£180**

Exercise

2 Copy and complete this table (give the answer to the nearest whole number where necessary).

Original cost	Saving	Fraction saved	Percentage saving
£32	£8	$\dfrac{£8}{£32} = \dfrac{1}{4} = \dfrac{25}{100}$	25%
£60	£6		
£14	£3		
£20	£3		
£21	£4		

3 Here are some advertisements for articles at reduced prices. In each case we have to calculate the *percentage reduction* in price (answer to nearest whole number).

Copy and complete the following:

(a) Original cost   = £72
Sale price   = £54
Reduction in cost = £18
Fraction saved   = $\dfrac{£18}{£72} \approx \dfrac{\blacksquare}{100} = \blacksquare\%$

ORIENTAL CARPETS
Also Rugs, in self-coloured embossed designs. Choice of three sizes, e.g.
2m × 1m, were £72
NOW _ _ _ _ _ _ _ _ _ **£54**

(b) Original cost   = £421
Sale price   = £325
Reduction in cost = £ 96
Fraction saved   = $\dfrac{£\ 96}{£421} \approx \dfrac{\blacksquare}{100} = \blacksquare\%$

"PAMPAS" SUITE
3- Pce. suite by 'Gimson Slater'
List Price £421
REDUCED TO _ _ _**£325**

4 Calculate the percentage reduction in price for each of the following articles (answer to nearest whole number).

JERSEY SHOP BARGAINS
2 and 3-Pce. suits, trouser suits, etc., e.g. "Lerose" dress. Was £20
NOW **£14**

LADIES' DRESSES
By "Pierre Cardin" Styled in pure wool. Sizes 12 - 16. Were £31.
REDUCED TO **£19**

BEDCOVERS
By "Casa Pupo" Double bed size in current colours. Originally £89. REDUCED TO **£39**

"SLUMBERLAND" DIVANS
Red Seal quality, firm edge divan. Size 5ft. x 6ft. 6in. Originally £188. REDUCED TO **£152**

HOSTESS TROLLEYS
The "Sovereign" by "Ekco" with large hot cupboard and heated top tray. List price £88.
OURS **£64**

PANTIE CORSELETTES
By "Au Fait" Sizes 34in. to 38in. bust. Originally £12.40.
REDUCED TO **£9**

Continue with Section P

P Profit and loss

A test of a good businessman is whether he makes a **profit** or a **loss**.
Consider the following two cases.

This man buys potatoes at the fruit market at 36p per kilo and sells them at 48p per kilo.

He makes a **profit**.

This man buys apples at 90p per kilo but has to cut his price to 72p per kilo to sell them.

He makes a **loss**.

It is not the actual size of the profit or loss that is important. A 1p profit on a 8p box of matches is a much better profit margin than a 1p profit on a 80p packet of cigarettes.

We can calculate the profit as a fraction of the cost price and write our answer as a **percentage profit.**

Exercise

1 Copy and complete this table (give percentages to nearest whole number where necessary)

Buying	Selling	Profit	Loss	Profit or loss as a percentage of buying price
2p	3p	1p	—	$\frac{1}{2} = \frac{50}{100} = 50\%$
20p	21p	1p	—	$\frac{1}{20} = \frac{}{100} = 5\%$
9p	12p			
30p	24p			
£175	£150			

2 A car which is bought for £800 is sold for £960. How much profit is made? What percentage is the profit of the buying price?

Copy and complete:

Buying price = £▧
Selling price = £▧
Profit   = £▧

$$\frac{\text{Profit}}{\text{Buying price}} = \frac{▧}{▧} = \frac{▧}{100} = ▧\%$$

The profit is ▧% of the buying price.

3 A man buys a house for £34 000 and sells it for £38 000. What percentage profit did he make?

4 A book which is bought for £1.80 is sold for £2.20. How much profit is this and what percentage is the profit of the buying price?

5 If a shopkeeper bought a refrigerator for £180 and sold it later for £140, how much did he lose on the deal?
What percentage is his loss of the buying price?

6 If you bought a bicycle for £52.80 and sold it later for £39, how much would you lose?
What percentage is your loss of the buying price?

In some industries, profit is calculated as a percentage of the selling price. In the following two questions calculate the profit as a percentage of the *selling price*.

7 A shopkeeper buys dog food at 28p per tin and sells it at 33p. Calculate the percentage profit.

Note *This time you have to calculate* $\dfrac{profit}{selling\ price}$.

8 Fish fingers are bought by the shopkeeper for 55p and sold for 69p. Calculate the percentage profit.

Continue with Section Q

Q Calculating the original amount

Example

This shirt has been increased in price by 30% and now costs £7.80. What was the price before the increase?

New price = original price + 30% of original price
         = 100% of original price + 30% of original price
         = 130% of original price

130% of original price = £7.80

So 1% of original price = $\frac{£7.80}{130}$ = 6p

100% of original price = 600p = £6

Original price = £6

Boys Casual Shirt with a shaped body. In a stone coloured Trevera/Cotton fabric Order sizes S. M. L.

Shirt £7.80

Exercise

1   The cost of an article is increased by 5%. The new price is £2.10. What was the original cost?

Example

This time we have a *reduction* in price. We have to find the original price of the anoraks.

Think of the original price as 100% and consider increases or decreases relative to this starting point of 100%.

A *decrease* in price (as in this example) of 10% means that the new price is 100% **minus** 10%, that is 90% of the original price.

90% of the original price   = £25.20

So 1% of the original price   = $\frac{2520p}{90}$ = 28p

So 100% of the original price = 2800p = £28

Original price = £28

sale

10% REDUCTION ON ALL REGULAR STOCK

MEN'S ANORAKS Available in a choice of sizes and colours. Ideal casual or sports wear. AT--- £25.20

Exercise

2   In a sale prices are reduced by 10%. What was the original price of a coat now costing £28.80?

3   A man's wages are increased by 8%. If he now earns £108 per week what did he earn previously?

4   A man has 25% of his pay deducted to pay his income tax. If his net salary (that is, his salary *after* paying income tax) is £7500, what is his gross salary (that is, his salary *before* tax is deducted)?

Continue with Section R

R Progress check

Exercise

1 A colour television set has a cash price of £360.

    (a) On HP terms the dealer wishes to make an extra 15% on the cash price. What must the dealers total HP charge amount to?

    (b) The deposit is 8% of the cash price. How much is the deposit?

    (c) Calculate the cost per month to repay the remaining amount over 24 equal monthly payments.

2 A salesman earns £110 per week, plus 3% commission on his sales. Calculate his wage in a week in which his sales were £1880.

EASY RUN

WHEELBARROW

From famous British Manufacturer — strongly constructed to take the heaviest load with 7/8″ green tubular steel frame and galvanized pan, 26″ x 21″ x 22¼″ with sides tapering to maximum 8½″ deep. Moves easily on 12″ diam. Solid tyre wheel. Value £27.

Save £6

Only £21

3 This wheelbarrow originally cost £27 and is being offered for £21. Calculate the percentage reduction in price (answer to nearest whole number).

4 A newspaper delivery boy has a paper round with weekly sales of £90. If he pays £72 for the newspapers, what is his percentage profit?

5 A shopkeeper buys computer games for £4.20 and sells them at £6.85. Calculate his percentage profit, giving your answer to the nearest whole number.

6 In a sale all articles are reduced by 15%. What was the original cost of a shirt whose sale price is £6.80?

7 Suncrest Computers Ltd. have increased the price of all their computers by 20%. The model Z computer now costs £180. What did it cost before the increase?

Tell your teacher you have finished this unit

UNIT 5 Probability

Probability introduced

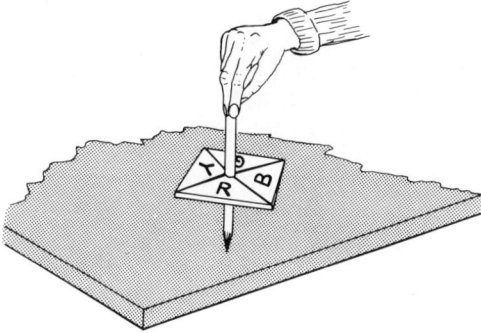

A square spinner has four equal sections coloured red, blue, green, and yellow.

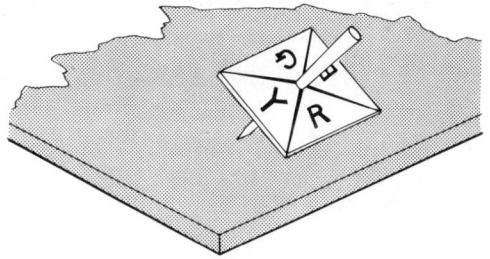

If we spin the spinner each coloured edge has the same chance of landing on the table.

So, for example, the red edge should land on the table once in every four spins.

We say that the **probability** of landing on red is $\frac{1}{4}$.

(Read $\frac{1}{4}$ as 'one in four' or as 'one quarter'.)

Example

What is the probability of landing on green?

> Number of green edges is 1.
> Number of edges is 4.
> Probability of landing on green is $\frac{1}{4}$.

Exercise

1 What is the probability of landing on blue?

Copy and complete: Number of blue edges is ▓.
Number of edges is ▓.

Probability of landing on blue is $\dfrac{▓}{▓}$.

2 What is the probability of *not* landing on blue?

Copy and complete: Number of edges *not* blue is ▓.
Number of edges is ▓.

Probability of *not* landing on blue is $\dfrac{▓}{▓}$.

3 What is the probability of landing on red or blue?

Copy and complete: Number of red and blue edges is ▓
Number of edges is ▓.

Probability of landing on red or blue is $\dfrac{2}{4} = \dfrac{1}{2}$.

4 Write an answer to the following in the same way as above. What is the probability of landing on yellow or green?

Continue with Section B

B | Dice

The die has six faces. If we throw the die each face has the same chance of coming up.

So, for example, the face scoring 2 should come up one in every six throws. We say that the probability of scoring 2 is $\frac{1}{6}$. (Read $\frac{1}{6}$ as 'one in six' or as 'one sixth'.)

Example

What is the probability of throwing a 3?

Number of faces marked 3 is 1.
Number of faces is 6.
Probability $= \frac{1}{6}$.

Exercise

1   What is the probability of *not* throwing a 3?

    Copy and complete: Number of faces *not* marked 3 is ▨.
                          Number of faces is ▨.

$$\text{Probability} = \frac{▨}{▨}.$$

Write answers to the following in the same way as above.

2   What is the probability of throwing a 5?

3   What is the probability of *not* throwing a 5?

4   What is the probability of throwing an even number?

5   What is the probability of throwing a number greater than 4?

    Copy and complete: Number of faces marked with numbers greater than 4 is ▨.
                          Number of faces is ▨.

$$\text{Probability} = \frac{2}{6} = \frac{1}{▨}.$$

6   What is the probability of throwing a number less than 4?

7   What is the probability of throwing a number divisible by 3?

8   What is the probability of throwing a number greater than 2?

Continue with Section C

C Cards

A pack of cards contains 52 cards which are divided into 4 suits.

Red Suits   **Black Suits**

  HEARTS   DIAMONDS   SPADES   CLUBS

Each suit contains 13 cards. In descending order they are:

Ace, King, Queen, Jack, 10, 9, 8, 7, 6, 5, 4, 3, 2.

Each card has the same chance of being drawn from a shuffled pack.

So, for example, the two of diamonds should be drawn once in every 52 draws from the full shuffled pack.

We say that the probability of drawing the two of diamonds is $\frac{1}{52}$.

We can write this as: Pr(two of diamonds) $= \frac{1}{52}$.

Exercise

1 What is the probability of drawing a black card?

Copy and complete: Number of black cards is ▨.

Number of cards in pack is ▨.

$$\text{Pr(black card)} = \frac{26}{52} = \frac{1}{2}.$$

2 What is the probability of drawing a face card (that is, King, Queen or Jack)?

Copy and complete: Number of face cards is ▨.

Number of cards in pack is ▨.

$$\text{Pr(face card)} = \frac{12}{52} = \frac{3}{13}.$$

3 Assuming you are drawing from a shuffled pack, copy and complete the following:

(a) $\text{Pr(red king)} = \dfrac{▨}{▨}$

(b) $\text{Pr(spade)} = \dfrac{▨}{▨}$

(c) $\text{Pr(black ace)} = \dfrac{▨}{▨}$

(d) $\text{Pr(a two)} = \dfrac{▨}{▨}$

(e) $\text{Pr(a diamond less than 7)} = \dfrac{▨}{▨}$ [*Do not count the ace.*]

(f) $\text{Pr(a club greater than 3)} = \dfrac{▨}{▨}$ [*Include the ace.*]

(g) $\text{Pr(a card less than 5)} = \dfrac{▨}{▨}$

(h) $\text{Pr(a card greater than 5)} = \dfrac{▨}{▨}$

Continue with Section D

D Probability scale

If we throw a die each face has the same chance of coming up.

The probability of any one of the faces coming up is $\frac{1}{6}$.

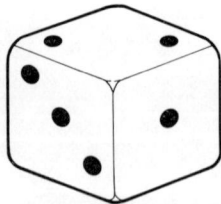

Example

What is the probability of scoring a 7?

No face is marked 7. Number of faces marked 7 is 0.

Number of faces on die is 6.

Pr (score of 7) $= \frac{0}{6} = 0$.

The probability of scoring a 7 is zero.

The probability of anything which is **impossible** is zero.

Example

What is the probability of scoring less than 7?

All six faces score less than 7. Number of faces scoring less than 7 is 6.
Number of faces on die is 6.
Pr(score less than 7) = $\frac{6}{6}$ = 1.

The probability of scoring less than 7 is 1.

> The probability of anything which is **certain** to happen is 1.

Exercise

Calculate the probabilities of the events listed below giving your answer as a decimal fraction:

1 Drawing a heart from a pack of cards.

2 Getting 'heads' with one toss of a coin.

3 Scoring more than zero with a die.

4 Not getting blue with the spinner in Section A.

5 Drawing a card less than two from a pack.

> The probability of any event cannot be less than zero or more than 1

Pr (drawing a heart) = 0.25

Certain event

1
0.9
0.8
0.7
0.6
0.5
0.4
0.3
0.2
0.1
0

Impossible event

6 Make a copy of the probability on the right and mark on the scale the probabilities calculated above.

7 Think of an event, work out its probability, and mark it on the Probability Wallchart.

Continue with Section E

E Expectation

Example

How many twos would we expect to get if a die is thrown 120 times?

$$\text{Pr(a two)} = \tfrac{1}{6}$$

Expected number of twos in 120 throws $= \tfrac{1}{6}$ of $120 = 20$.

Exercise

1 What is the expected number of sixes in 12000 throws of an unbiased die?

2 What is the expected number of odd number scores in 1000 throws of a fair die?

3 How many heads would you expect to get if a coin is tossed 500 times?

4 What is the expected number of scores less than 3 in 3000 throws of a fair die?

Example

2 out of 5 people use *Sowhite* toothpaste. If 100 people are interviewed how many of them will be expected to use *Sowhite*?

$$\text{Pr(a person using } Sowhite) = \tfrac{2}{5}$$

Expected number in 1000 interviews $= \tfrac{2}{5}$ of $1000 = 2 \times 200 = 400$.

Exercise

5 3 out of 5 housewives use *Brash* soap powder. If 500 housewives are interviewed how many are expected to use *Brash*?

6 5 out of 8 car–owners have a *Forall* car. If 1600 car–owners are interviewed how many are expected to own a *Forall*?

7 An opinion poll predicts that 4 out of 5 voters will vote for Mr Brown. How many votes might Mr Brown expect if 2500 voted?

8 1 in 3 people read the *Daily News*. If 750 people are questioned, how many might be expected to read the *Daily News*?

Continue with Section F

F Two dice

Here are two dice.

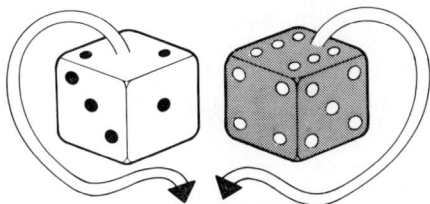

We can show the score made by the dice like this ———————————————→ (2, 6)

Let us arrange all the possible pairs of numbers using the two dice.

Exercise

Copy and complete the following table. Remember that the number on the white die always appears first.

(1, 1)	(1, 2)	(1, 3)	(1, 4)	(1, 5)	(1, 6)
(2, 1)	(2, 2)	(2, 3)	(2, 4)	(2, 5)	
(3, 1)	(3, 2)	(3, 3)	(3, 4)		
(4, 1)	(4, 2)	(4, 3)			
(5, 1)	(5, 2)				
(6, 1)					

There are 36 different pairs.

Example

What is the probability of the first number being a 1?

(1, 1)	(1, 2)	(1, 3)	(1, 4)	(1, 5)	(1, 6)
(2, 1)	(2, 2)	(2, 3)	(2, 4)	(2, 5)	(2, 6)
(3, 1)					

Number of pairs with 1 in the first place is 6.
Number of pairs is 36.
Pr(a 1 in first place) $= \frac{6}{36} = \frac{1}{6}$.

Exercise

1   What is the probability of the second number being a 3?

Copy and complete the following:

Number of pairs with second number 3 is ▨.

Number of pairs is ▨.

Pr(second number a 3) $= \frac{▨}{▨} = \frac{▨}{▨}$.

(1, 2)	(1, 3)	(1, 4)
(2, 2)	(2, 3)	(2, 4)
(3, 2)	(3, 3)	(3, 4)
(4, 2)	(4, 3)	(4,
(5, 2)	(5, 3)	(5, 4
(6, 2)	(6, 3)	(6, 4)

2 What is the probability of the first number being greater than the second?

Copy and complete:

(2, 1)					
(3, 1)	(3, 2)				
(4, 1)	(4, 2)	(4, 3)			
(5, 1)	(5, 2)	(5, 3)	(5, 4)		
(6, 1)	(6, 2)	(6, 3)	(6, 4)	(6, 5)	(6, 6)

Number of pairs with first number greater than the second is ▨.
Number of pairs is ▨.

$$\text{Pr(first greater than second)} = \frac{▨}{▨} = \frac{▨}{▨}.$$

3 What is the probability of the second number being greater than the first?

Copy and complete:

Number of pairs with second number greater than the first is ▨. Number of pairs is ▨.

$$\text{Pr(second greater than first)} = \frac{▨}{▨} = \frac{▨}{▨}.$$

(1, 1)	(1, 2)	(1, 3)	(1, 4)	(1, 5)	(1, 6)
(2, 2)	(2, 3)	(2, 4)	(2, 5)	(2, 6)	
(3, 3)	(3, 4)	(3, 5)	(3, 6)		
(4, 4)	(4, 5)	(4, 6)			
(5, 5)	(5, 6)				

Continue with Section G

G Problems with two dice

Your finished table of pairs in Section F should look like this.

(1, 1)	(1, 2)	(1, 3)	(1, 4)	(1, 5)	(1, 6)
(2, 1)	(2, 2)	(2, 3)	(2, 4)	(2, 5)	(2, 6)
(3, 1)	(3, 2)	(3, 3)	(3, 4)	(3, 5)	(3, 6)
(4, 1)	(4, 2)	(4, 3)	(4, 4)	(4, 5)	(4, 6)
(5, 1)	(5, 2)	(5, 3)	(5, 4)	(5, 5)	(5, 6)
(6, 1)	(6, 2)	(6, 3)	(6, 4)	(6, 5)	(6, 6)

Example

What is the probability of throwing a double? [*That is, of getting the same score on both dice.*]

Number of doubles is 6. [(1, 1), (2,2), (3,3), (4,4), (5,5), (6,6)]
Number of pairs is 36.
Pr(a double) $= \frac{6}{36} = \frac{1}{6}$.

Exercise

1 What is the probability of scoring a double six? [*That is, getting (6, 6).*]

2 Find the probability of scoring a total of 2. [*That is, adding the scores of the two dice gives 2.*]

3 Find the probability of scoring a total of 6.

4 What is the probability of getting a total score which is an even number?

Example

What is the probability of scoring
a total of 8 or more?

(1, 1)	(1, 2)	(1, 3)	(1, 4)	(1, 5)	(1, 6)
(2, 1)	(2, 2)	(2, 3)	(2, 4)	(2, 5)	(2, 6)
(3, 1)	(3, 2)	(3, 3)	(3, 4)	(3, 5)	(3, 6)
(4, 1)	(4, 2)	(4, 3)	(4, 4)	(4, 5)	(4, 6)
(5, 1)	(5, 2)	(5, 3)	(5, 4)	(5, 5)	(5, 6)
(6, 1)	(6, 2)	(6, 3)	(6, 4)	(6, 5)	(6, 6)

Number of pairs totalling 8 or more
is 15.
Number of pairs is 36.
Pr(total 8 or more) $= \frac{15}{36} = \frac{5}{12}$.

Exercise

5 Find the probability of scoring a total of 5 or less.

6 Find the probability of scoring at least 6. ['*At least 6' means that the totals could be 6, 7, 8, 9, 10, 11 or 12.*]

7 Find the probability of scoring a total of 10 or more.

8 What is the probability of scoring a total less than 10?

Example

What is the probability of scoring a total of 1?

Number of pairs totalling 1 = 0.
Number of pairs = 36.
Pr(total of 1) $= \frac{0}{36} = 0$.

It is impossible to score a total of 1, so the probability equals zero.

Exercise

9 What is the probability of scoring a total greater than 1, but less than 13?

Continue with Section H

H Two spinners

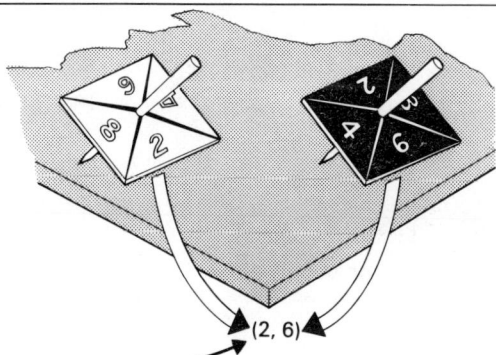

Here are two spinners. One is white, the
other black.

We can show the score made by the
spinners like this (2, 6)

Exercise

1 Copy the table opposite and complete it to
show all the possible pairs of numbers
using the two spinners. Remember that
the number on the white spinner always
appears first.

(2, 2)	(2,)	(2,)	(2,)
(4,)	(,)	(,)	(,)
(6,)	(,)	(,)	(,)
(8,)	(,)	(,)	(,)

Your finished table should look like this

Exercise

2 Lay your pencil over

 (a) the pairs which total 8,
 (b) the pairs which total 10,
 (c) the pairs which total 12.

(2, 2)	(2, 4)	(2, 6)	(2, 8)
(4, 2)	(4, 4)	(4, 6)	(4, 8)
(6, 2)	(6, 4)	(6, 6)	(6, 8)
(8, 2)	(8, 4)	(8, 6)	(8, 8)

Example

What is the probability of
scoring a total of 8?

(2, 2)	(2, 4)	(2, 6)	(2, 8)
(4, 2)	(4, 4)	(4, 6)	(4, 8)
(6, 2)	(6, 4)	(6, 6)	(6, 8)
(8, 2)	(8, 4)	(8, 6)	(8, 8)

$$\text{Pr(total of 8)} = \frac{\text{Number of pairs scoring 8}}{\text{Total number of pairs}}$$

$$= \frac{3}{16}$$

Exercise

3 What is the probability of scoring a total of 10?

 Copy and complete:

$$\text{Pr(total of 10)} = \frac{\text{Number of pairs scoring 10}}{\text{Total number of pairs}} = \frac{\blacksquare}{\blacksquare} = \frac{\blacksquare}{\blacksquare}$$

4 What is the probability of scoring a total of 12?

 Copy and complete:

$$\text{Pr(total of 12)} = \frac{\text{Number of pairs scoring 12}}{\text{Total number of pairs}} = \frac{\blacksquare}{\blacksquare}$$

5 What is the probability of scoring a double?

 Copy and complete:

$$\text{Pr(a double)} = \frac{\text{Number of doubles}}{\text{Total number of pairs}} = \frac{\blacksquare}{\blacksquare} = \frac{\blacksquare}{\blacksquare}$$

6 What is the probability of scoring an odd total?

 Copy and complete:

$$\text{Pr(odd total)} = \frac{\text{Number of odd totals}}{\text{Total number of pairs}} = \frac{\blacksquare}{\blacksquare} = \frac{\blacksquare}{\blacksquare}$$

7 What is the probability of scoring an even total?

 Copy and complete:

$$\text{Pr(even total)} = \frac{\text{Number of even totals}}{\text{Total number of pairs}} = \frac{\blacksquare}{\blacksquare} = \frac{\blacksquare}{\blacksquare}$$

Continue with Section I

A die and a coin

Here we have a die and a coin.

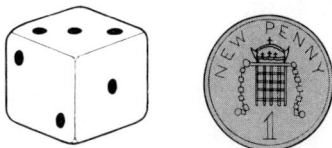

We can show the results of throwing the
die and tossing the coin like this_____ (3, T)

Exercise

1 Copy the table below and complete it to show all the possible pairs using a die and a coin.
The number on the die is written first.

(1, H)	(,)	(3, H)	(,)	(,)	(,)
(,)	(2, T)	(,)	(,)	(,)	(6, T)

2 What is the probability of getting a 5 and a 'tail'?

Copy and complete:

$$\text{Pr}(5, \text{tail}) = \frac{\text{Number of pairs with 5 and 'tail'}}{\text{Total number of pairs}} = \frac{\blacksquare}{\blacksquare}$$

3 What is the probability of getting an even number and a 'head'?

Copy and complete:

$$\text{Pr}(\text{even, head}) = \frac{\text{Number of pairs with even number and 'head'}}{\text{Total number of pairs}} = \frac{\blacksquare}{\blacksquare} = \frac{\blacksquare}{\blacksquare}$$

4 What is the probability of getting a number less than 5 and a 'tail'?

Copy and complete:

$$\text{Pr}(\text{less than 5, tail}) = \frac{\text{Number of pairs with number less than 5 and tail}}{\text{Total number of pairs}} = \frac{\blacksquare}{\blacksquare} = \frac{\blacksquare}{\blacksquare}$$

Continue with Section J

J Tree diagrams

We can show the possible results of throwing a die and tossing a coin in a diagram.

List six numbers on the die ↓	Write H and T against each number ↓	List all the possible results alongside ↓
1	H / T	(1, H) / (1, T)
2	H / T	(2, H) / (2, T)
3	H / T	(3, H) / (3, T)
4	H / T	(4, H) / (4, T)
5	H / T	(5, H) / (5, T)
6	H / T	(6, H) / (6, T)

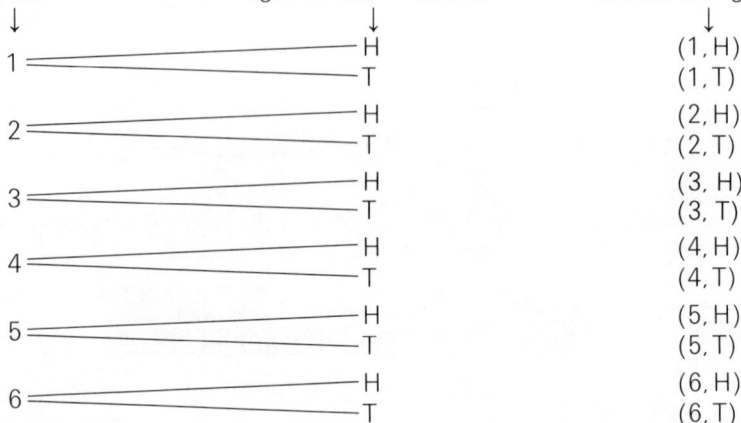

Example

What is the probability of throwing a 5 with a 'tail'?

There is one result with a 5 and a 'tail'.
There are 12 possible results.

$\text{Pr}(5, \text{tail}) = \frac{1}{12}$

Example

Find Pr(not 6, tail).

There are 5 possible ways of not getting a 6 with a tail.
$[(1, T), (2, T), (3, T), (4, T), (5, T).]$
There are 12 possible results.
$\text{Pr}(\text{not } 6, \text{tail}) = \frac{5}{12}$

Exercise

1 Copy and complete this diagram and the list of possible results for spinning the square spinner and tossing the coin.

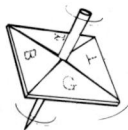

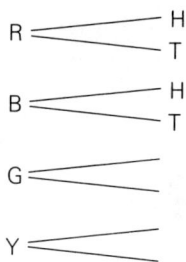

Possible Results

R $\diagdown$ H   (R, H)
      T   (R, T)

B $\diagdown$ H   (,)
      T   (,)

G $\diagdown$ (,)
            (,)

Y $\diagdown$ (,)
            (,)

2 (a) What is the probability of landing on a red edge and getting a 'head'?

Copy and complete:

$$\text{Pr(red edge, head)} = \frac{\text{Number of pairs with R and H}}{\text{Total number of pairs}} = \frac{\blacksquare}{\blacksquare}$$

(b) Find Pr(*not* green edge, tail).

Continue with Section K

K Multiplication of probabilities

We found in Section J that the probability of getting (5, tail) is $\frac{1}{12}$.

The probability of getting is $\frac{1}{6}$. That is, Pr(5) $= \frac{1}{6}$.

The probability of getting is $\frac{1}{2}$. That is, Pr(tail) $= \frac{1}{2}$.

Now $\frac{1}{6} \times \frac{1}{2} = \frac{1}{12}$,
which is the probability of getting

That is Pr(5, tail) $=$ Pr(5) × Pr(tail).

You found in Section J that the probability of getting (red, tail) is $\frac{1}{8}$.

The probability of getting is $\frac{1}{4}$. That is, Pr(red) $= \frac{1}{4}$.

The probability of getting is $\frac{1}{2}$. That is, Pr(tail) $= \frac{1}{2}$.

Now $\frac{1}{4} \times \frac{1}{2} = \frac{1}{8}$,
which is the probability of getting

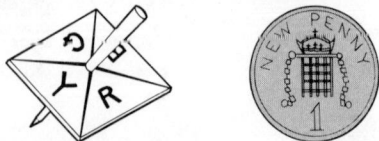

So Pr(red, tail) $=$ Pr(red) × Pr(tail)

Exercise

Find the probabilities of the following:

Copy and complete this column.

1  drawn from a complete pack.

$Pr(\text{three of diamonds}) = \dfrac{\blacksquare}{\blacksquare}$

2  with one toss of a coin.

$Pr(\text{head}) = \dfrac{\blacksquare}{\blacksquare}$

3  with one throw of a die.

$Pr(3) = \dfrac{\blacksquare}{\blacksquare}$

4 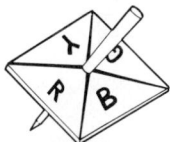 with one spin of the spinner.

$Pr(\text{blue}) = \dfrac{\blacksquare}{\blacksquare}$

5

$Pr(\text{head, three of diamonds}) = \dfrac{\blacksquare}{\blacksquare} \times \dfrac{\blacksquare}{\blacksquare} = \dfrac{\blacksquare}{\blacksquare}$

6

$Pr(3,\text{blue}) = \dfrac{\blacksquare}{\blacksquare} \times \dfrac{\blacksquare}{\blacksquare} = \dfrac{\blacksquare}{\blacksquare}$

7

$Pr(\text{blue, three of diamonds}) = \dfrac{\blacksquare}{\blacksquare} \times \dfrac{\blacksquare}{\blacksquare} = \dfrac{\blacksquare}{\blacksquare}$

8

$Pr(\text{head},3) = \dfrac{\blacksquare}{\blacksquare} \times \dfrac{\blacksquare}{\blacksquare} = \dfrac{\blacksquare}{\blacksquare}$

9

$Pr(3,3) = \dfrac{\blacksquare}{\blacksquare} \times \dfrac{\blacksquare}{\blacksquare} = \dfrac{\blacksquare}{\blacksquare}$

10

$Pr(\text{head, head}) = \dfrac{\blacksquare}{\blacksquare} \times \dfrac{\blacksquare}{\blacksquare} = \dfrac{\blacksquare}{\blacksquare}$

Continue with Section L

L Beads in a box

In the box there are 3 red beads and 2 blue beads. The beads are identical except for colour.

The box is shaken and a bead is taken out without looking.

What is the probability that it is a red bead?

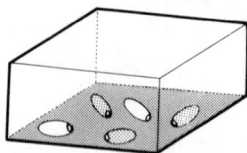

$$Pr(red) = \frac{\text{Number of red beads}}{\text{Total number of beads}} = \frac{3}{5}$$

What is the probability of taking out a red bead, putting it back in the box, and again taking out a red bead?

$$Pr(red, red) = Pr(red) \times Pr(red) = \frac{3}{5} \times \frac{3}{5} = \frac{9}{25}$$

Exercise

Using the method shown above, calculate the following:

1 The probability of taking out a red bead followed by a blue bead.
That is Pr(red, blue).
[*Remember that the first bead is put back in the box before taking out the second.*]

2 The proability of taking out a blue bead followed by a red bead.
That is Pr(blue, red).

3 The probability of taking out a blue bead followed by a blue bead.
That is Pr(blue, blue).

4 In this bag there are 1 white and 4 red billiard balls.

The bag is shaken and a ball is taken out.

(a) What is the probability that it is red?

(b) If this ball is put back into the bag and again a ball is taken out, what is the probability that it will be white?

(c) What is the probability of taking a red ball, replacing it in the bag, and then taking a white ball?

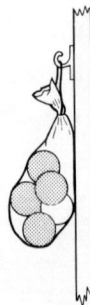

Continue with Section M

M Three spinners

A regular ten–sided spinner has 2 green sides, 3 blue sides, and 5 red sides.

The spinner is spun and the colour of the side on which it falls is noted.

Pr(R)   is the probability of spinning a red.

Pr(R, G)   is the probability of spinning red then green.

Pr(R, G, B) is the probability of spinning red then green then blue.

Exercise

Copy and complete the probabilities below.

1 Pr(R) $= \dfrac{5}{10} = \dfrac{1}{\blacksquare}$

2 Pr(B) $=$

3 Pr(G) $=$

4 Pr(R, R) $= Pr(R) \times Pr(R) = \dfrac{1}{2} \times \dfrac{1}{2} = \dfrac{1}{\blacksquare}$

5 Pr(R, B) $=$

6 Pr(B, G) $=$

7 Pr(G, G) $=$

8 Pr(R, G) $=$

9 Pr(B, R, G) $= Pr(B) \times Pr(R) \times Pr(G) = \dfrac{3}{10} \times \dfrac{1}{2} \times \dfrac{1}{5} = \dfrac{\blacksquare}{100}$

10 Pr(R, R, R) $=$

11 Pr(B, B, B) $=$

12 Pr(G, G, G) $=$

Continue with Section N

N Progress check

Exercise

1 A die is thrown. What is the probability of getting

(a) a six, (b) an even number, (c) a number divisible by 3?

2 A triangular spinner is spun.
Calculate Pr(red).

3 One of the letters of the word POPOCATEPETL is chosen at random. What is the probability that it is

(a) P, (b) C, (c) T, (d) a letter other than P?

4 (a) Write down an event whose probability is 0.
(b) Write down an event whose probability is 1.

5 (a) How many sixes would you expect if a die is thrown 3000 times?
(b) How many reds would you expect if the spinner in question 2 is spun 3000 times?
(c) 2 out of 3 people have TV in their homes. If 3000 people are interviewed, how many can be expected to have TV in their homes?

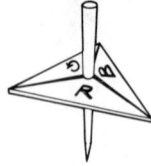

6 A die and a coin are tossed. Write down all the possible results. What is the probability that the die shows

(a) a three, (b) a number less than three,
(c) a number greater than 2 and the coin shows a head?

7 Two dice are thrown. What is the probability that

(a) the dice will land (1,1), (b) the total score is even,
(c) the total score is ten, (d) the total score will be less than 5,
(e) one die at least shows a six?

8 A bag contains 1 red, 2 black, and 3 white marbles. A marble is taken from the bag without looking, is replaced, and again a marble is taken. Calculate

(a) Pr(R,R), (b) Pr(R,B), (c) Pr(B,W),
(d) Pr(B,B).

Ask your teacher what to do next

O Success and failure

In a jar there are 100 beads of which three
are marked with a black cross.

A bead is taken at random from the jar.

Probability that it is marked $= \dfrac{3}{100}$,

$\quad$ Pr(marked) $= \dfrac{3}{100}$.

Probability that it is not marked $= \dfrac{97}{100}$,

$\quad$ Pr(not marked) $= \dfrac{97}{100}$.

So, Pr(marked) + Pr(not marked) $= \dfrac{3}{100} + \dfrac{97}{100} = \dfrac{100}{100} = 1$.

Example

In an experiment it was discovered that the probability of a seed germinating was $\frac{7}{10}$.
What is the probability that the seed will not germinate?

$$Pr(G) + Pr(\text{not } G) = 1$$
$$\text{So } \tfrac{7}{10} + Pr(\text{not } G) = 1$$
$$\text{So } Pr(\text{not } G) \quad = \tfrac{3}{10}$$

In statistics it is usual to talk about the results of trials as **success** or **failure**.

In the above example if the seed germinating is called **success** and the seed not
germinating is called **failure** [*that is 'failure' means 'not success'*] then

$$Pr(\text{Success}) = \frac{7}{10} \text{ and } Pr(\text{Failure}) = \frac{3}{10}$$

So $\qquad Pr(S) + Pr(F) = \dfrac{7}{10} + \dfrac{3}{10} = 1$.

Exercise

Write answers to the following.

1 A card is drawn at random from a well–shuffled pack, What is the probability that it is
$\quad$ (a) a spade, (b) not a spade?

2 $\frac{1}{10}$ of the bolts made by a machine are faulty. What are the probabilities that a bolt picked
at random from the machine's output will be
$\quad$ (a) faulty, (b) not faulty?

3 0.06 of children write with the left hand. What is the probability that a child, picked at
random, will
$\quad$ (a) write with the left hand,
$\quad$ (b) not write with the left hand?

4 One person in every five takes tea without sugar. If success means choosing a person at random who takes tea without sugar, what is the probability of

(a) success, (b) failure?

5 The probability of an insecticide being effective is 0.9, that is, Pr(Success) = 0.9. What is Pr(Failure)?

Continue with Section P

P Multiplication of probabilities

Example

Two dice are thrown. What is the probability that a six appears on the first die but not on the second?

$$\Pr(6) = \frac{1}{6} \qquad \Pr(\text{not } 6) = 1 - \frac{1}{6} = \frac{5}{6}$$

$$\Pr(6, \text{not } 6) = \frac{1}{6} \times \frac{5}{6} = \frac{5}{36}$$

Example

A five sided spinner is spun four times. What is the probability that a 5 does not result from the first three spins, but does result from the fourth?

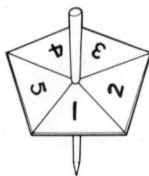

That is, what is Pr(not 5, not 5, not 5, 5)?

$$\Pr(5) = \frac{1}{5} \qquad \Pr(\text{not } 5) = 1 - \frac{1}{5} = \frac{4}{5}$$

$$\text{So } \Pr(\text{not } 5, \text{not } 5, \text{not } 5, 5) = \frac{4}{5} \times \frac{4}{5} \times \frac{4}{5} \times \frac{1}{5}$$

$$= \frac{4 \times 4 \times 4 \times 1}{625}$$

$$= \frac{64}{625}$$

Exercise

1 Three pennies are tossed.

(a) What is the probability that the three coins show heads?
(b) What is the probability that the three coins show tails?
(c) What is the probability that the first two coins show tails and the third shows a head?

2 A die is thrown three times.

(a) What is the probability that a four turns up on each of the three throws? [*Find Pr(4,4,4).*]
(b) What is the probability that a four turns up on the first throw but not on the other two throws? [*Find Pr(4, not 4, not 4).*]
(c) What is the probability that on the first throw a three turns up, on the second a six turns up, and on the third throw a two turns up? [*Find Pr(3, 6, 2).*]

3 A box contains 10 beads, three of which are red and the rest black. A bead is taken from the box three times. (The bead is put back in the box before another is taken.)

(a) What is the probability that the first two beads are red and the third black? [*Find Pr(R,R,B).*]
(b) Find Pr(R, B, B)   (c) Find Pr(R, R, R)   (d) Find Pr(B, B, B)

4 Four pennies are tossed. Find the following probabilities.

(a) Pr(H, H, H, H)   (b) Pr(H, H, T, T)   (c) Pr(T, T, T, T)

5 Four dice are thrown. Find the following probabilities.

(a) Pr(1, 1, 1, 1)   (b) Pr(1, not 1, 1, not 1)   (c) Pr(6, 6, 6, 6)

6 In an experiment the probability of success in one trial is found to be $\frac{1}{3}$. That is, Pr(S) = $\frac{1}{3}$. Find the following probabilities.

(a) Pr(S)   (b) Pr(F)   (c) Pr(S, S)   (d) Pr(S, F)
(e) Pr(F, F)   (f) Pr(F, S)   (g) Pr(S, S, S)   (h) Pr(S, S, F)
(i) Pr(S, F, F)   (j) Pr(F, F, S)   (k) Pr(F, S, F)   (l) Pr(F, F, F)

Continue with Section Q

Q Number of arrangements

Example

Some competitions require that a number of photographs be put in a certain order to win a prize.

A

B

C

How many different arrangements are there of the photographs A, B, and C?

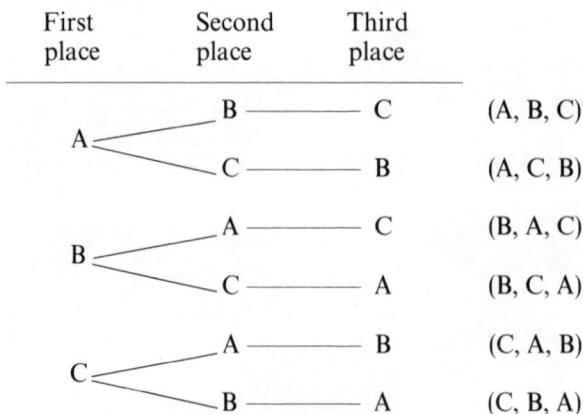

First place	Second place	Third place	
A	B	C	(A, B, C)
	C	B	(A, C, B)
B	A	C	(B, A, C)
	C	A	(B, C, A)
C	A	B	(C, A, B)
	B	A	(C, B, A)

There are six different arrangements and only one is correct.

Probability of having the correct order = $\frac{1}{6}$.

Example

How many different arrangements are there of the photographs A, B, C, and D?
When A is in the first place there are 6 different arrangements as shown below:

First place	Second place	Third place	Fourth place	
	B	C	D	(A, B, C, D)
		D	C	(A, B, D, C)
	C	B	D	(A, C, B, D)
A		D	B	(A, C, D, B)
	D	C	B	(A, D, C, B)
		B	C	(A, D, B, C)

Exercise

1 Draw a similar diagram with B in first place.

2 How many arrangements are there with B in first place?

3 How many arrangements are there with C in first place?

4 How many arrangements are there with D in first place?

5 How many different arrangements are there altogether?

6 What is the probability of choosing one of these arrangements?

Continue with Section R

R Progress check

Exercise

1 In a multiple–choice question there are five possible answers given, only one of which is correct. By guessing, what is the probability of choosing
(a) the correct answer, (b) one of the wrong answers.

2 A pack of cards is cut. The pack is then shuffled and cut again. What is the probability of cutting a 'diamond' first and 'not a diamond' second?

3 Two dice are thrown at the same time. Find the following probability:

(a) Pr(3, not 3),   (b) Pr(not 3, 3),   (c) Pr(not 3, not 3).

4 Numbers 7, 8, and 9 were marked on slips of paper. The three slips were rolled up and placed in a hat. Three–digit numbers are formed by drawing the numbers one at a time from the hat. Write out all the possible three–digit numbers. What is the probability of drawing 879 from the hat?

> **Tell your teacher you have finished this unit**

UNIT 6 Formulae

A Day of birth

On which day of the week were you born?

It is possible using a formula to find out which day of the week you were born on. In this Unit we will find out the days of birth of Daley Thompson, David Bowie, the Duke of Edinburgh, Sebastian Coe, Indira Gandhi and Margaret Thatcher.

The following formula will enable you to find the day of the week for any date this century:

$$\text{Day of the week} = \text{Remainder of } \frac{D+M+Y}{7}$$

where D = Day of the month (1 to 31),
M = Month key number, and
Y = Year key number.

Here are the tables we use.

Month key number, M						
0	1	2	3	4	5	6
January (*leap year*)	January		**February** (*leap year*)	February		
				March		
April		May			June	
July			August			September
	October			November		December

Year key number, Y						
0	1	2	3	4	5	6
	1901	1902	1903		**1904**	1905
1906	1907		**1908**	1909	1910	1911
	1912	1913	1914	1915		**1916**
1917	1918	1919		**1920**	1921	1922
1923		**1924**	1925	1926	1927	
1928	1929	1930	1931		**1932**	1933
1934	1935		**1936**	1937	1938	1939
	1940	1941	1942	1943		**1944**
1945	1946	1947		**1948**	1949	1950
1951		**1952**	1953	1954	1955	
1956	1957	1958	1959		**1960**	1961
1962	1963		**1964**	1965	1966	1967
	1968	1969	1970	1971		**1972**
1973	1974	1975		**1976**	1977	1978
1979		**1980**	1981	1982	1983	
1984	1985	1986	1987		**1988**	1989
1990	1991		**1992**	1993	1994	1995
	1996	1997	1998	1999		**2000**

Remainder
0 = Saturday
1 = Sunday
2 = Monday
3 = Tuesday
4 = Wednesday
5 = Thursday
6 = Friday

Note Leap years are marked in heavy black print.

Example

On which day of the week was Paul McCartney born? His date of birth is 18th June, 1942.

$$\frac{D+M+Y}{7} = \frac{18+5+3}{7}$$

$$= \frac{26}{7}$$

$$= 3 \text{ remainder } 5$$

So Paul McCartney was born on a Thursday.

Example

On which day of the week was Linda McCartney born? Her date of birth is 24th September, 1941.

$$\frac{D+M+Y}{7} = \frac{24+6+2}{7}$$

$$= \frac{32}{7}$$

$$= 4 \text{ remainder } 4$$

So Linda McCartney was born on a Wednesday.

Remainder
0 = Saturday
1 = Sunday
2 = Monday
3 = Tuesday
4 = Wednesday
5 = Thursday
6 = Friday

Exercise

1   Which day of the week is it today? Use the formula to check the day.

2   On which day of the week were you born?

3   Here is a list of famous people and their birth dates. Choose any six and find out on which day of the week they were born:

(a) *Film Stars*   Robert De Niro (17th August 1943); Meryl Streep (22nd June 1949); Jeremy Irons (19th September 1948); Ben Kingsley (31st December 1943); Dustin Hoffman (8th August 1937); Natasha Kinski (24th January 1961).

(b) *Television and Pop Stars*   Terry Wogan (3rd August 1948); Ronnie Corbett (4th December 1930); Cilla Black (27th May 1943); Lulu (3rd November 1948); David Bowie (8th January 1947).

(c) *Royal Family*   Duke of Edinburgh (10th June 1921); Queen Elizabeth (21st April 1926); Prince Charles (14th November 1948); Princess Diana (1st July 1961); Princess Anne (15th August 1950).

(d) *Politicians*   Neil Kinnock (28th March 1942); Margaret Thatcher (13th October 1925); Ronald Reagan (6th February 1911).

(e) *Sports Stars*   Tessa Sanderson (14th March 1956); Daley Thompson (30th July 1958); Glen Hoddle (27th March 1957); Seve Ballesteros (9th April 1957); Sebastian Coe (29th September 1956).

Continue with Section B

B Triangular display

How many tins are on display?

The formula for finding the number, N, of tins in a triangular display is as follows:

$$N = \frac{n(n+1)}{2}$$

where n = number of tins at the base.

Example

How many tins are in the triangular display? (There are 9 tins at the base.)

$$N = \frac{n(n+1)}{2} \quad \text{where } n = 9$$

$$= \frac{9(9+1)}{2}$$

$$= \frac{9 \times 10}{2}$$

$$= \frac{90}{2}$$

$$= 45$$

There are 45 tins in the display.

Checking this answer by counting the number of tins in each row and adding we have:
$N = 9+8+7+6+5+4+3+2+1 = 45$.

Exercise

1 How many tins are in the triangular display shown below. (7 tins at the base)

Copy and complete: $\quad N = \frac{n(n+1)}{2} \quad$ where $n = 7$.

$$= \frac{7(7+1)}{2}$$

$$= \frac{7 \times \blacksquare}{2}$$

$$= \frac{\blacksquare}{2}$$

$$= \blacksquare$$

There are $\blacksquare$ tins in the display.
By counting, number of tins = $\blacksquare$

2 Here is a pyramid of a different kind.

It uses the same formula where n means the number of boys on the *top* row and N means the total number of boys.

Copy and complete: $N = \dfrac{n(n+1)}{2}$ where $n = 3$.

$$= \frac{\rule{2cm}{0.3cm}}{2}$$

$$= \frac{\rule{1cm}{0.3cm}}{2}$$

$$= \rule{1cm}{0.3cm}$$

There would be ▨ boys altogether.

3 Suppose we could have 10 boys on the top row, how many boys would there be in the whole pyramid?

Copy and complete: $N = \dfrac{n(n+1)}{2}$ where $n = $ ▨

$$= \frac{\rule{2cm}{0.3cm}}{2}$$

$$= \frac{\rule{1cm}{0.3cm}}{2}$$

$$= \rule{1cm}{0.3cm}$$

There would be ▨ boys altogether. (Please don't try it out.)

Continue with Section C

C Sectional area

Sometimes we have to find the areas of curved shapes, or unusual shapes.

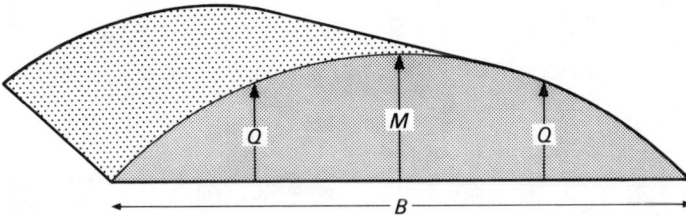

Here is a formula for finding the **area** of a shape like the *end* of the hangar shown in the diagram.

$$\text{Area} = \frac{B(4Q+M)}{6}$$

where B = breadth,
     M = height of the middle, and
     Q = height quarter of the way along.

The area found is called the **cross-sectional area** of the hangar, that is, the area of the shape which fits all the way along its length.

Example

Find the area of the end of the hangar if its measurements are as shown below.

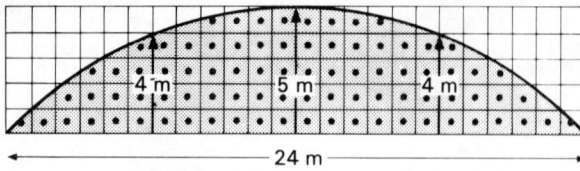

24 m

$$A = \frac{B(4Q+M)}{6}$$

$$= \frac{24(4 \times 4 + 5)}{6}$$

where $B = 24$,
$M = 5$, and
$Q = 4$.

$$= \frac{24(16+5)}{6} = \frac{24 \times 21}{6} = \frac{504}{6} = 84$$

That is, the area is 84 m².

Exercise

1   Check, by counting the squares in the diagram which are marked with a dot, that this answer is reasonable.

2   Find the cross–sectional area of the hangar below.

Copy and complete:

$$A = \frac{B(4Q+M)}{6}$$

where $B = 18$, $Q = 5$, and $M = 6$.

$$= \frac{18(4 \times 5 + 6)}{6}$$

$$= \frac{18(\blacksquare + 6)}{6}$$

$$= \frac{18 \times \blacksquare}{6}$$

$$= \frac{\blacksquare}{6}$$

$$= \blacksquare$$

The cross–sectional area is ▇ m².

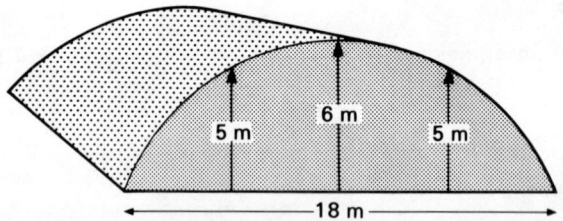

5 m   6 m   5 m

18 m

3   The drawing shows the shape and measurements of a fireside rug. Find its area.

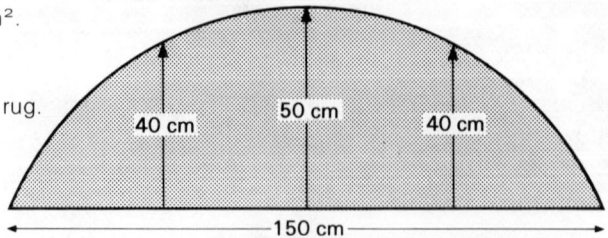

40 cm   50 cm   40 cm

150 cm

4   Find the area of a plot of ground with shape and measurements as shown.

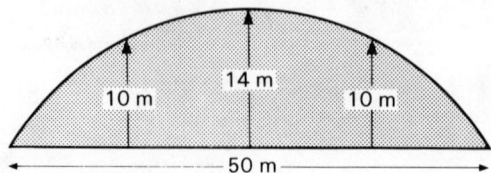

10 m   14 m   10 m

50 m

Continue with Section D

D Length

The examples you have met so far have shown some formulae in action. Here is a list of some formulae which are commonly used to calculate **lengths or distances**.

1 Perimeter of a triangle: $P = a+b+c$

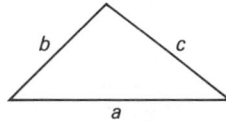

2 Perimeter of a rectangle: $P = 2(l+b)$

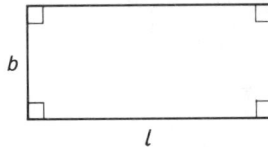

3 Circumference of a circle: $C = \pi d$ ($\pi \approx 3.14$)

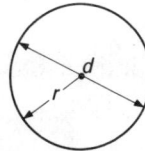

4 Length of longest side of a right-angled triangle: $c^2 = a^2+b^2$

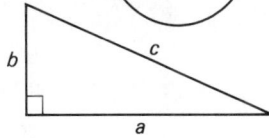

Example

Find the perimeter of a triangle with sides 13 cm, 7 cm, and 9 cm.

$P = a+b+c$ where $a = 13$, $b = 7$, and $c = 9$.
$= 13+7+9$
$= 29$

Perimeter $= 29$ cm.

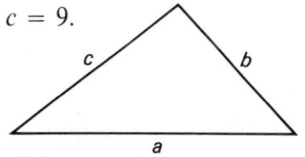

Exercise

Write answers in the same way as in the example above.

1 Find the perimeter of a triangle whose sides are
(a) 7.62 cm, 8.53 cm, 5.26 cm,   (b) 14.2 cm, 11.8 cm, 8.5 cm.

Example

Find the perimeter of a rectangle of length 17.2 cm and breadth 9.4 cm.

$P = 2(l+b)$ where $l = 17.2$ and $b = 9.4$.
$= 2(17.2+9.4)$
$= 2 \times 26.6$
$= 53.2$

Perimeter $= 53.2$ cm.

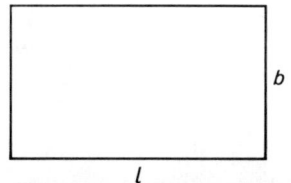

Exercise

Write answers in the same way as in the example.

2   Find the perimeter of a rectangle which has sides of
   (a) length 115 m and breadth 63 m,
   (b) length 16.6 cm and breadth 12.7 cm.

Example

Find the circumference of a circle with diameter 7 cm.

$$C = \pi d \quad \text{where } d = 7.$$
$$= 3.14 \times 7$$
$$= 21.98$$
$$\text{Circumference} = 21.98 \text{ cm.}$$

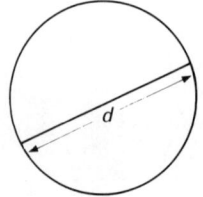

Exercise

Write answers in the same way as in the example.

3   Find the circumference of a circle with

   (a) diameter 30 cm,   (b) diameter 12 cm,   (c) radius 4 cm.

Example

Find the length of the hypotenuse of a right-angled triangle whose other sides are 5.4 cm and 6.7 cm.

$$c^2 = a^2 + b^2 \quad \text{where } a = 5.4 \text{ and } b = 6.7.$$
$$= 5.4^2 + 6.7^2$$
$$= 29.16 + 44.89 \quad \boxed{\text{Table of squares are useful here.}}$$
$$= 74.05$$
$$c = \sqrt{74.05} = 8.61$$
$$\text{Hypotenuse} = 8.61 \text{ cm.}$$

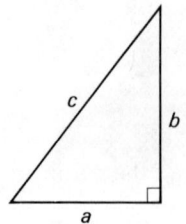

Exercise

Write answers in the same way as in the example.

4   Find the hypotenuse of a right angled triangle with other sides
   (a) 4 cm and 7 cm,   (b) 2.5 cm and 4.1 cm.

Continue with Section E

E Area

Here are some formulae used to calculate **areas**.

1   Area of a rectangle: $A = lb$

2   Area of a circle: $A = \pi r^2$ ($\pi \approx 3.14$)

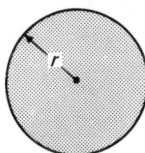

3   The curved surface of a cylinder can be thought of as a rectangle whose length is the same as the circumference of the circular ends.

Curved surface area of a cylinder: $A = \pi dh$
where d = diameter and h = height ($\pi \approx 3.14$).

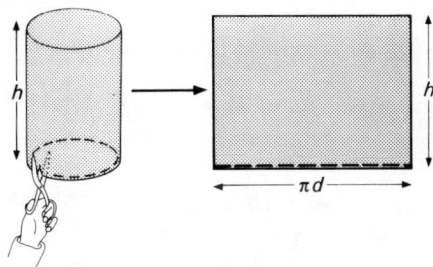

4   Area of a triangle: $A = \frac{1}{2}bh$
where b = length of base and h = height.

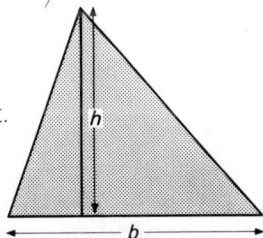

Example

Calculate the area of a rectangle of length 4 cm and breadth 3 cm.

$A = lb$ where l = 4 and b = 3.
$$= 4 \times 3$$
$$= 12$$
Area = 12 cm^2.

Exercise

Write answers in the same way as in the example.

1   Calculate the area of a rectangle which has sides of
(a) length 8 m and breadth 7 m,   (b) length 4.6 cm and breadth 2.7 cm.

Example

Calculate the area of a circle which has radius 8 cm.

$$A = \pi r^2 \quad \text{where } r = 8.$$
$$= 3.14 \times 8 \times 8$$
$$= 3.14 \times 64$$
$$= 200.96$$
$$\text{Area} = 200.96 \, \text{cm}^2.$$

Exercise

Write answers in the same way as in the example.

2 Calculate the area of a circle which has

    (a) radius 5 cm,   (b) radius 30 m,   (c) diameter 22 cm.

Example

Calculate the curved surface area of a cylinder of height 12 cm and radius 5 cm.

$$A = \pi d h \quad \text{where } h = 12 \text{ and } d = 10.$$
$$= 3.14 \times 10 \times 12$$
$$= 31.4 \times 12$$
$$= 376.8$$
$$\text{The curved surface area} = 376.8 \, \text{cm}^2.$$

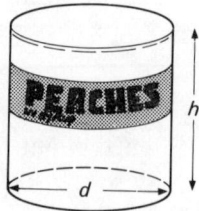

Exercise

Write answers in the same way as in the example.

3 Calculate the curved surface area of a cylinder with

    (a) height 10 cm and radius 5 cm,   (b) height 25 cm and radius 2 cm.

Example

Calculate the area of a triangle with base 14 cm and height 12 cm.

$$A = \tfrac{1}{2} b h \quad \text{where } b = 14 \text{ and } h = 12.$$
$$= \tfrac{1}{2} \times 14 \times 12$$
$$= 7 \times 12$$
$$= 84$$
$$\text{Area} = 84 \, \text{cm}^2.$$

Exercise

Write answers in the same way as in the example.

4 Calculate the area of a triangle with

    (a) base 24 cm and height 18 cm,   (b) base 48 cm and height 100 cm.

Continue with Section F

F Areas of compound shapes

Exercise

1 A dining room has a rectangular floor 4.2 m long and 3.7 m broad. What is the area of the floor?

2 Calculate the area of a shape with the sizes given in the diagram.

Hint: Calculate the area of parts A and B and then add them.

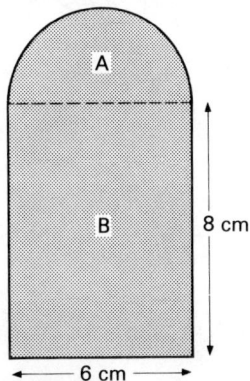

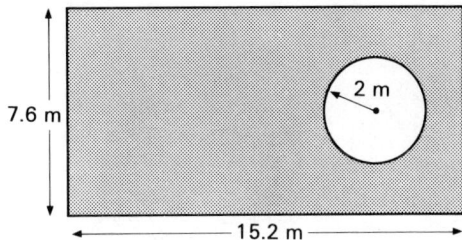

3 Find the area of lawn shown in the diagram.

Hint: Calculate the area of the rectangle and the area of the circle and then subtract.

4 Find the area of the washer shown in the drawing.

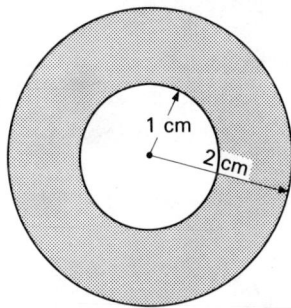

5 Calculate the area of this shape.

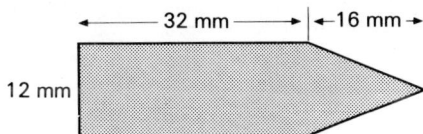

6 Calculate the area of a figure with these dimensions.

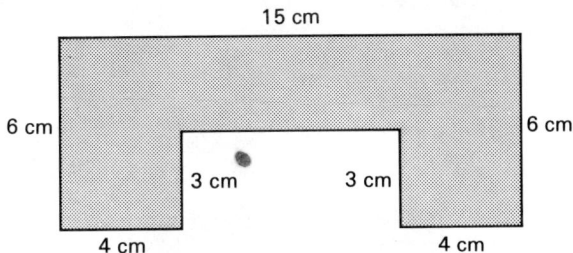

Continue with Section G

G Volume

The formulae in this section are used to calculate **volumes.**

1   Volume of a cuboid: $V = lbh$

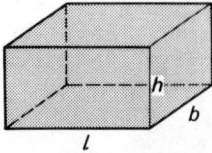

2   Volume of a cube: $V = l^3$

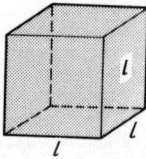

3   Volume of a cylinder:

$$V = \pi r^2 h$$

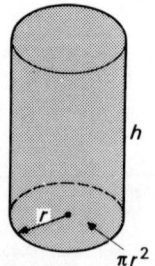

Example

Find the volume of a cylinder whose base radius is 10 cm and whose height is 30 cm.

$V = \pi r^2 h$ where $\pi = 3.14$, $r = 10$, and $h = 30$.

$= 3.14 \times 10^2 \times 30$

$= 3.14 \times 100 \times 30$

$= 314 \times 30$

$= 9420$

The volume of the cylinder is 9420 cm³.

Exercise

Use the formulae given above in the following questions.

1   Find the volume of a cuboid with

(a) length 2 cm, breadth 4 cm, height 5 cm,
(b) length 1.7 cm, breadth 3 cm, height 10 cm.

2   Find the volume of a cube with

(a) length 8 cm,   (b) length 2.5 cm.

3   Find the volume of a cylinder with

(a) base radius 5 cm, height 16 cm,
(b) base radius 6 cm, height 10 cm.

4 Find the volume of metal used to make a cylindrical pipe whose inner radius is 30 cm and outer radius is 40 cm. The length of the pipe is 5 metres.

Hint: Find the volume of the outer cylinder and the volume of the inner cylinder and then subtract.

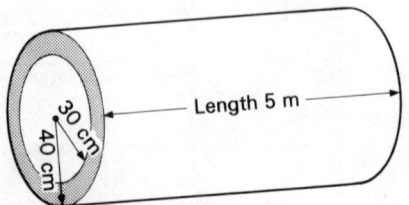

Continue with Section H

H Prisms

A shape which has the same cross section from top to bottom is called a **prism.**

Here are some examples.

Cubes and cuboids are **prisms.**

To find the volume of a prism we need to find the area of the cross section and multiply by the height.

That is, the formula is $V = Ah.$

Note: The cross–sectional area may be at the end of the prism. In this case we can call the other dimension the *length* of the prism and use the formula $V = Al.$

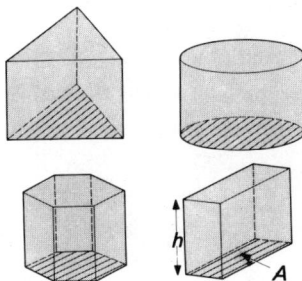

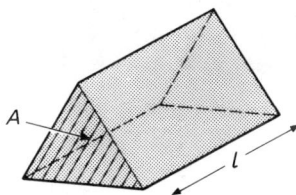

Example

The hangar shown in the diagram on page 106 has a cross-sectional area of $84\,m^2$. Find its volume if its length is $20\,m$.

$$V = Al$$
$$= 84 \times 20 \quad \text{where the cross sectional area is } A = 84,$$
$$= 1680 \qquad\qquad \text{and the length is } l = 20.$$
Volume $= 1680\,m^3$.

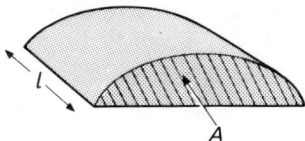

Exercise

1 Find the volumes of the prisms shown below.

(a)

11.0 cm

Area 6 cm²

(b)

Area 8 m²

4 m

2 A packet of cheese is shaped like this. Find the volume of cheese which it holds.

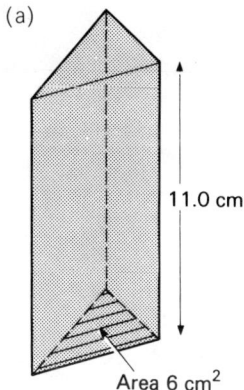

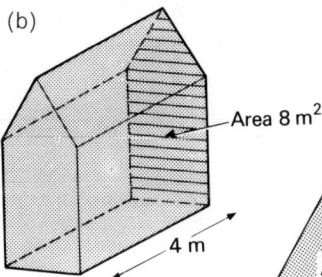

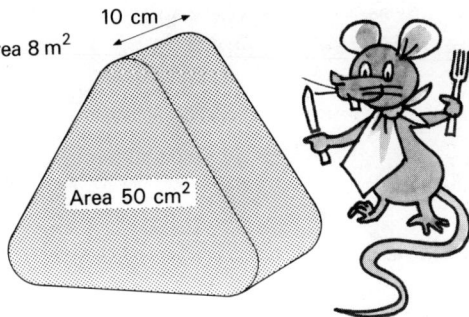

10 cm

Area 50 cm²

I Substitution in formulae

Example

The volume of a prism is given by the formula $V = Ah$.

Suppose we know that the volume is $850 \, cm^3$ and the area of the cross section is $50 \, cm^2$. What is the height?

Use the formula

$$V = Ah$$
$$850 = 50h$$
$$\frac{850}{50} = h \qquad \text{Divide both sides by 50.}$$
$$17 = h$$

So height of prism is 17 cm.

Exercise

1 A rectangle has an area of 350 m². If its length is 25 m, find its breadth. (Use $A = lb$.)

2 A rectangle has a perimeter of 80 m. If its length is 30 m, find its breadth. (Use $P = 2l + 2b$.)

3 A fish tank has a base with area 1500 cm².
   If 6 litres of water are poured in, how deep
   will the water be?
   (1 litre = 1000 cm³.)

   Use Volume = Area × depth,
   that is, $V = Ad$.

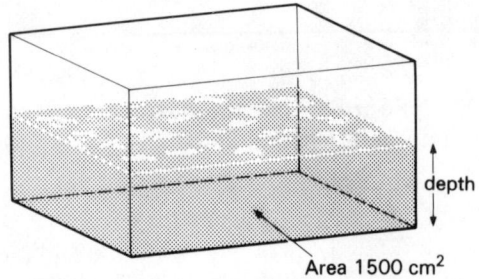

depth

Area 1500 cm²

4 A rectangular garage base measures 8 m
   long and 5 m wide. If 2 m³ of concrete are
   poured in, how deep will the base be?

   (Use $V = lbh$.)

8 m

5 m

depth

5 Find the diameter of a circle whose circumference is 50 cm.

   (Use $C = \pi d$) (Give answer correct to nearest centimetre.)

6 Find the radius of a circle whose area is 100 cm².

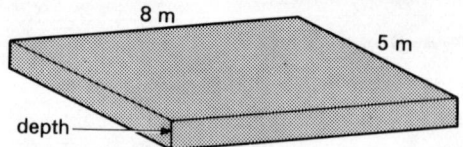

   Copy and complete the following:

   $$A = \pi r^2$$
   $$100 = 3.14 \times r^2$$
   $$\frac{100}{3.14} = r^2$$
   $$\text{So } r^2 = \blacksquare$$
   $$r = \sqrt{\blacksquare} = \blacksquare$$
   So radius is $\blacksquare$ cm.

7 Find the radius of a circle whose area is 60 cm².

Continue with Section J

J Progress check

Here is a list of formulae for **distance**, **area**, and **volume**.

Distance

Perimeter of triangle:	$p = a + b + c$
Perimeter of rectangle:	$P = 2l + 2b$
Circumference of circle:	$C = \pi d$ or $C = 2\pi r$
Theorem of Pythagoras:	$c^2 = a^2 + b^2$ or $c = \sqrt{a^2 + b^2}$

Area

Area of triangle:	$A = \frac{1}{2}bh$
Area of rectangle:	$A = lb$
Area of circle:	$A = \pi r^2$
Curved surface area of cylinder:	$A = \pi dh$ or $A = 2\pi rh$

Volume

Volume of cuboid:	$V = lbh$
Volume of prism:	$V = Ah$
Volume of cylinder:	$V = \pi r^2 h$

Exercise

1 A hangar has a cross section with breadth, B, of 25 m, middle height, M, of 8 m, and quarter height, Q of 7 m. Find its area.

Use $A = \dfrac{B(4Q + M)}{6}$.

2 What is the circumference of a circle with radius 8 cm?

3 Find the length of the hypotenuse of a right-angled triangle whose shorter sides are 4.7 cm and 6.2 cm respectively.

4 A cylinder has a radius of 5 cm and a height of 8 cm. Calculate

    (a) its curved surface area,   (b) its volume.

5 Find the areas of the following figures:

1.7 cm

2.8 cm

4.6 cm

1.4 m

1.4 m

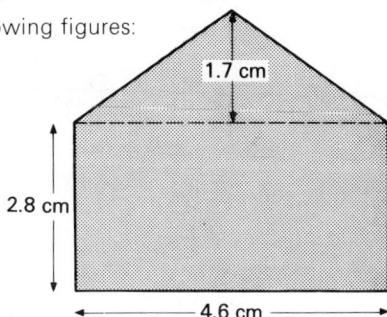

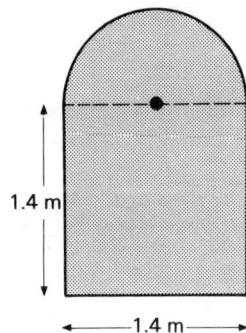

6 What is the diameter of a circle whose circumference is 30.2 cm? (Give answer correct to one decimal place.)

7 A water tank is a cuboid in shape. It contains 108 m³ of water. If the length of the tank is 8 m and the breadth is 5 m, what is the height of the water?

Ask your teacher what to do next

K Flowcharts

When we use a formula we are given certain numbers and using the formula we calculate our result.

For example given values for the length and breadth of a rectangle we can calculate the area. ($A = lb$)

In computer terminology, given certain **input** numbers we can **output** a result.

This may be shown in a **flowchart.**

Example

The flowchart below shows the method of calculating the monthly salary of a salesman who is paid a basic salary plus a commission on sales.

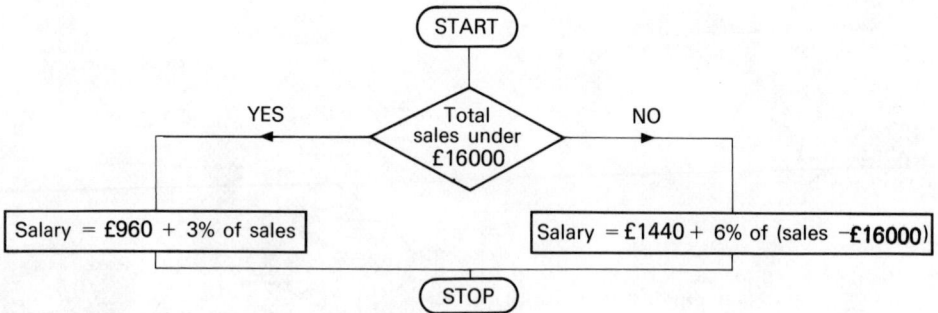

```
                        ( START )
                            |
         YES            Total              NO
      <--------  <  sales under  >  -------->
      |              £16000             |
      |                  |              |
 +-------------------+        +-----------------------------------+
 | Salary = £960 + 3% of sales |   | Salary = £1440 + 6% of (sales -£16000) |
 +-------------------+        +-----------------------------------+
                            |
                        ( STOP )
```

What are the salaries payable to salesmen with total sales of (a) £14000, (b) £18800?

(a) £14000
In this case we follow the YES route
Salary $= £960 + 3\%$ of £14000
$= £960 + 3 \times £140$
$= £960 + £420$
$= £1380$

(b) £18800
In this case we follow the NO route
Salary $= £1440 + 6\%$ of $(18\,800 - 16\,000)$
$= £1440 + 6\%$ of £2800
$= £1440 + 6 \times £28$
$= £1440 + £168$
$= £1608$

Exercise

1 Using the flowchart above calculate the monthly salary payable to a salesman with sales of (a) £10000, (b) £20000.

2 The flowchart below shows how salaries are calculated in a different firm.

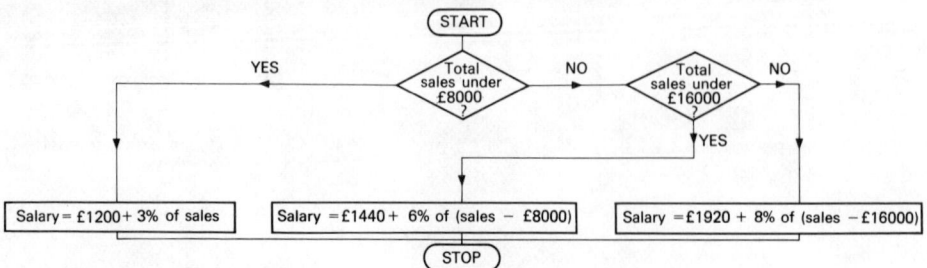

```
                              ( START )
                                  |
       YES             Total            NO            Total            NO
    <--------    <  sales under  >  -------->   <  sales under  >  -------->
    |                 £8000                          £16000             |
    |                   ?                              ?                |
    |                   |                              | YES            |
    |                   |                              |                |
+------------------+ +--------------------------+ +-----------------------------------+
| Salary = £1200 + 3% of sales | | Salary = £1440 + 6% of (sales - £8000) | | Salary = £1920 + 8% of (sales - £16000) |
+------------------+ +--------------------------+ +-----------------------------------+
                                  |
                              ( STOP )
```

What are the salaries payable to salesmen with total sales of
(a) £6000,   (b) £13 000,   (c) £20 600?

3 The cost of posting second class letters depends on the weight and can be calculated using the following flowchart.

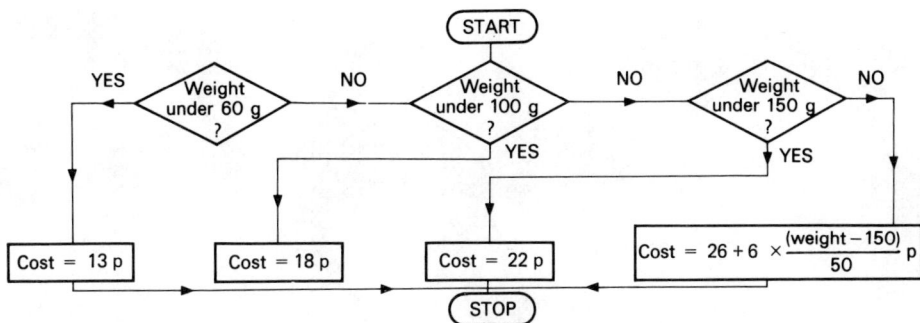

What is the cost of posting letters weighing
(a) 50 g,   (b) 90 g,   (c) 140 g,   (d) 250 g,   (e) 500 g

4 A shop offers a discount to customers depending on the total cost of their bill. The details are shown in the following flowchart.

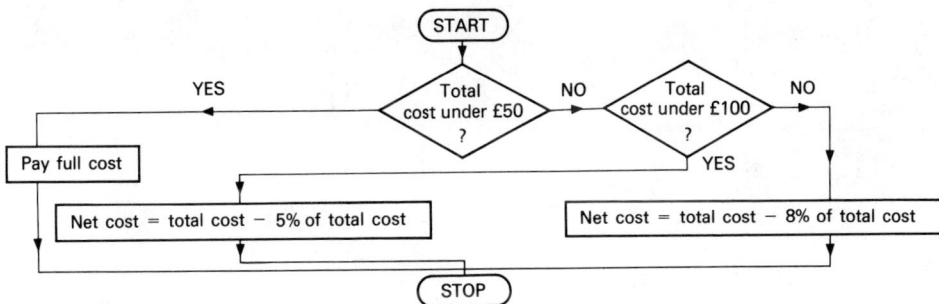

Find the net cost paid by customers whose purchases had a total cost of
(a) £36.50,   (b) £90,   (c) £325.

Continue with Section L

L Area of curved surface

Two formulae for curved surfaces are given below.

1 Curved surface area of a cone:

$$A = \pi rs$$ where s is the slant height.

2 Curved surface area of a sphere:

$$A = 4\pi r^2$$

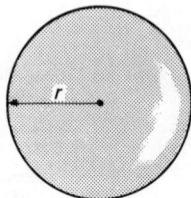

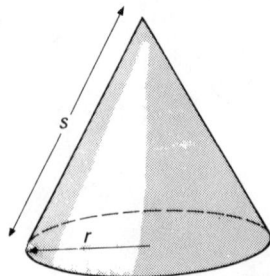

Example

Find, to the nearest square centimetre, the curved surface area of a cone whose base has radius 5 cm and whose slant height is 13 cm. What is its total surface area in square centimetres to the nearest ten?

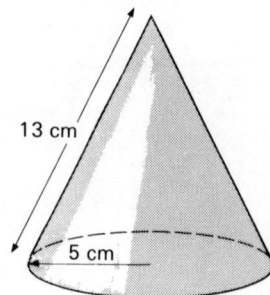

13 cm

5 cm

Area of Curved Surface	*Area of Base*
$A = \pi rs$	$A = \pi r^2$
$= 3.14 \times 5 \times 13$	$= 3.14 \times 5^2$
≈ 204.1	$= 3.14 \times 25$
	$= 78.5$

Area $= 204.1\,\text{cm}^2$

Total surface area $= 204.1 + 78.5 = 282.6 \approx 280\,\text{cm}^2$ (to the nearest ten).

Exercise

1 Find, to the nearest square centimetre, the curved surface area of the cones with following dimensions

(a) base radius 5.2 cm, slant height 8.8 cm,
(b) base radius 2.3 cm, slant height 4.6 cm.

2 Find, to the nearest square centimetre, the surface area of the spheres with following dimensions

(a) radius 4 cm,   (b) diameter 1.9 cm.

3 A boiler, in the shape of a cylinder with hemispherical ends, has a total length of 5 m and its diameter is 2 m. Find the cost (to the nearest ten pounds) of the sheet metal used to make it if it costs £48 per m².

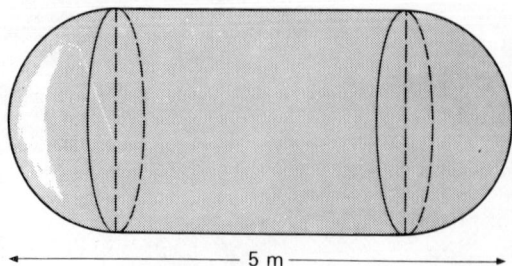

5 m

4 A buoy is constructed from a hemisphere topped by a cone. The diameter of the hemisphere is 1.6 m and the slant height of the cone is 2 m. If the steel sheeting from which it is made weighs 35 kg/m², find the weight of the buoy to the nearest kilogram.

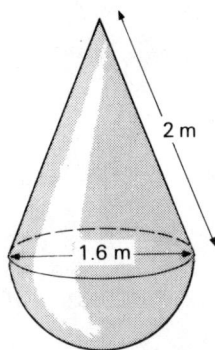

2 m

1.6 m

Continue with Section M

M Volume of pyramid, cone, and sphere

Any formula is like a recipe. It shows how to use the various parts to produce the result.

Here are some more formulae for calculating volumes.

Volume of a pyramid: $V = \frac{1}{3}Ah$

Volume of a cone: $V = \frac{1}{3}\pi r^2 h$

Volume of a sphere: $V = \frac{4}{3}\pi r^3$

Example

Find the volume of a pyramid with a base area of 30 cm² and a height of 12 cm.

$V = \frac{1}{3}Ah$ where $A = 30$ and $h = 12$.
$= \frac{1}{3} \times 30 \times 12$
$= 10 \times 12$
$= 120$

Volume of pyramid $= 120\,\text{cm}^3$.

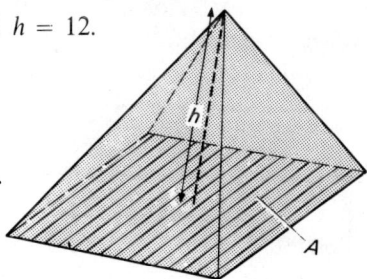

We can use any formula in this way, even although we haven't seen it before.

Example

Find the volume (to the nearest ten cubic centimetres) of a sphere with radius 10 cm.

$V = \frac{4}{3}\pi r^3$ where $r = 10$.
$= \frac{4}{3} \times 3.14 \times 10 \times 10 \times 10$
$= \frac{4}{3} \times 3140$
$= \dfrac{12560}{3}$
≈ 4190

Volume of sphere $= 4190\,\text{cm}^2$.

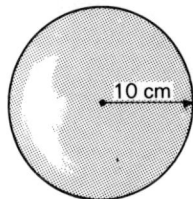

Exercise

1   Find the volume, to the nearest cubic centimetre, of a rectangular pyramid of height 7.62 cm, and base 4.2 cm by 6.5 cm. (Note that the base of the pyramid is a *rectangle* with Area = 4.2 × 6.5.)

2   A pyramid 8.75 m high stands on a square base of side 3.5 m. Find its volume to the nearest cubic metre. (Note that the base of the pyramid is a square with Area = 3.5 × 3.5.)

3   A pyramid stands on an octagonal base which has an area 28.1 cm² and the height of the pyramid is 5.5 cm. What is its volume to the nearest cubic centimetre?

4   Find the volume, to the nearest cubic centimetre, of a cone with

   (a) base radius 3 cm, height 5 cm,
   (b) base radius 2.5 cm, height 4.15 cm,
   (c) base diameter 3.5 cm, height 2.6 cm.

5   Find the volume, to the nearest cubic centimetre, of a sphere with

   (a) radius 3 cm,   (b) radius 4.3 cm,   (c) diameter 9.3 cm.

6   Find the weight, to the nearest gram, of a glass marble 2.3 cm in diameter, if 1 cubic centimetre of glass weighs 2.7 g.

Continue with Section N

N Volumes of compound shapes

Very often we are given questions in which two or more shapes are combined, or in which the given dimensions are not exactly those used in the formulae. In most of these questions a good diagram will help you to understand what is required.

Exercise

In each of the following questions copy the diagram.

1   A solid figure is formed from a cylinder with a cone on top as shown.

   (a) If the cylinder has radius 14 cm and height 28 cm, find the volume of the cylinder.
   (b) If the cone has height 12 cm, what is the volume of the cone?
   (c) What is the volume of the complete shape to the nearest thousand cubic centimetres?

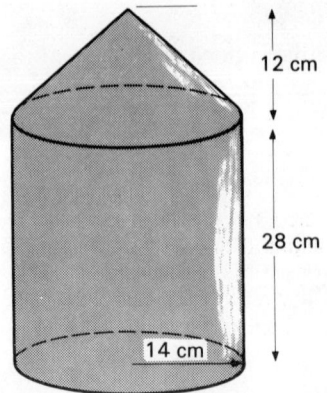

12 cm

28 cm

14 cm

2   What is the volume of the largest sphere that can be put into a cubical box of edge 24 cm? (Answer to the nearest hundred cubic centimetres.)

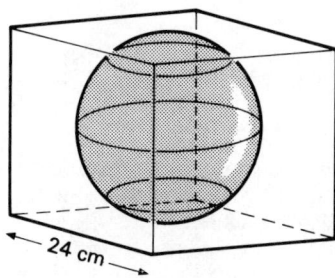

← 24 cm →

3   The ice–cream cone shown in the diagram is filled with ice cream and topped with a hemisphere of ice cream. Calculate the volume of ice cream used, to the nearest ten cubic centimetres.

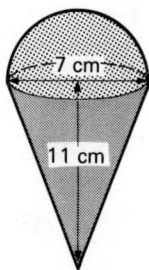

7 cm

11 cm

4   A petrol storage tank is in the shape of a cylinder with hemispherical ends. Its total length is 5 m and its diameter is 2 m. If petrol weighs 750 kg/m³, what weight of petrol can it hold when full? (Answer to the nearest ten kilograms.)

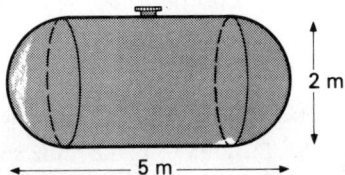

2 m

← 5 m →

Continue with Section O

O Problems

In addition to the formulae on page 115, we have now used the following:

Area

Curved surface area of a cone: $A = \pi rs$

Surface area of a sphere: $A = 4\pi r^2$

Volume

Volume of a pyramid: $V = \frac{1}{3}Ah$

Volume of a cone: $V = \frac{1}{3}\pi r^2 h$

Volume of a sphere: $V = \frac{4}{3}\pi r^3$

Exercise

Using the above formulae answer the following questions:

1   Calculate the volume of a pyramid of height 6 cm and length and breadth as shown in the diagram.
(Answer to the nearest cubic centimetre.)

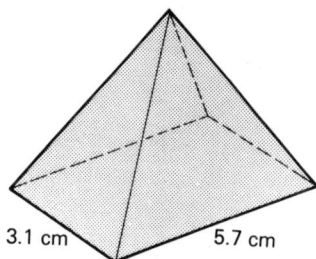

3.1 cm   5.7 cm

2 A wooden top is in the shape of a cone with slant height 6 cm together with a hemisphere of diameter 6 cm. Calculate the total surface area of the top to the nearest square centimetre.

3 Steel ball bearings are manufactured with a diameter of 0.6 cm. Calculate the volume of 1000 ball bearings. If 1 cm³ of steel weighs 7.8 g, what would be the weight of these ball bearings? (Answer to the nearest ten grams.)

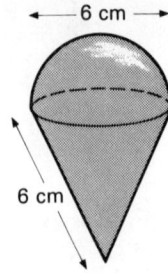

Continue with Section P

P Change of subject

Sometimes it is easier and quicker to change the subject of a formula before any calculations are begun.

This is especially true if you are using the same formula for a series of calculations.

Example

The formula for the area of a triangle is $A = \frac{1}{2}bh$.

Change this formula to make h the subject of the formula.

$$A = \tfrac{1}{2}bh$$

Multiply both sides by 2.

$$2A = bh$$

Divide both sides by b.

$$\frac{2A}{b} = h$$

So

$$h = \frac{2A}{b}$$

Exercise

Change the subject of each formula to the letter shown:

1 (a) $C = \pi d$ (to d) (b) $V = lbh$ (to h) (c) $V = \frac{1}{3}Ah$ (to h)

Example

Change the subject of the formula $P = 2(l+b)$ to l.

$$P = 2(l+b)$$

Divide both sides by 2.

$$\frac{P}{2} = l+b$$

Subtract b from both sides.

$$\frac{P}{2} - b = l$$

So

$$l = \frac{P}{2} - b$$

Exercise

Change the subject of each formula.

2 (a) $A = 3(x+y)$ (to y)   (b) $A = 2h(l+b)$ (to b)
   (c) $A = 2\pi r(r+h)$ (to h)

Example

Change the subject of the formula $V = Ah^2$ to h.

$$V = Ah^2$$

Divide both sides by A.

$$\frac{V}{A} = h^2$$

So

$$h^2 = \frac{V}{A}$$

Take the square root of both sides.

$$h = \sqrt{\frac{V}{A}}$$

Exercise

Change the subject of each formula to r.

3 (a) $V = \pi r^2 h$ (b) $A = 4\pi r^2$ (c) $V = \frac{1}{3}\pi r^2 h$

Example

Change the subject of the formula $T = 2\pi\sqrt{\dfrac{l}{g}}$ to l.

$$T = 2\pi\sqrt{\frac{l}{g}}$$

Square both sides.

$$T^2 = \frac{4\pi^2 l}{g}$$

Multiply both sides by g.

$$T^2 g = 4\pi^2 l$$

Divide both sides by $4\pi^2$.

$$\frac{T^2 g}{4\pi^2} = l$$

So

$$l = \frac{T^2 g}{4\pi^2}$$

Exercise

Change the subject of each formula.

4 (a) $k = V\sqrt{d}$ (to d)   (b) $d = \sqrt{b^2 - 4ac}$ (to a)
   (c) $c = \sqrt{2hR - h^2}$ (to R)

5 $T = \dfrac{P}{r^2}$ (to P)   **6** $V = d^2\sqrt{H}$ (to d)

7 $d = \dfrac{k-m}{t}$ (to k)   **8** $c = \dfrac{Prn}{100}$ (to r)

9 $r = px+q$ (to x)   **10** $q = 1 + \dfrac{2}{p}$ (to p)

Continue with Section Q

Q Progress check

Exercise

1 A holiday firm offer reduced rates for party bookings. The following flowchart shows their charges.

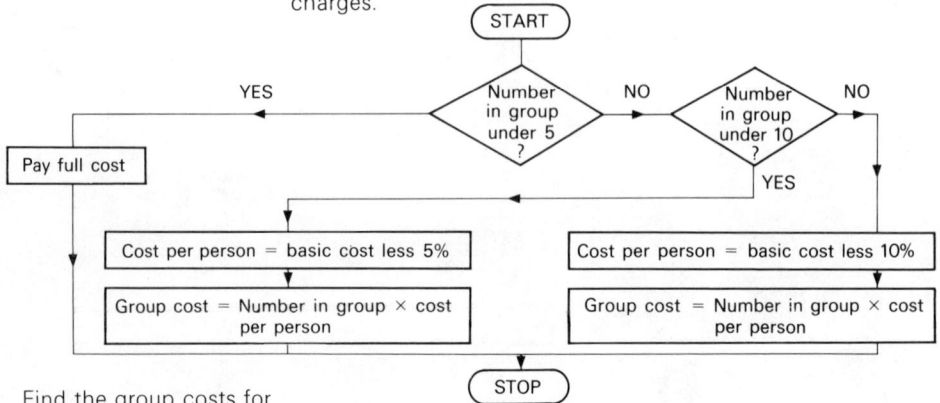

Find the group costs for
(a) 4 people – basic cost £280 per person. (b) 8 people – basic cost £480 per person,
(c) 12 people – basic cost £640 per person.

2 Calculate the volume of wood in a hemispherical bowl with internal diameter 20 cm and external diameter 22 cm.

3 Find the weight of a stone monument consisting of a pyramid 30 m high on a square base with edge 4 m and standing on a cubical block with edge 4 m, if 1 m³ of stone weighs 10 kg.

4 A sports hall is built in the form of a hemisphere with diameter 30 m. The whole surface is to be given a protective coating of plastic at a cost of 20p per square metre. Calculate the cost of this treatment.

5 A large bell tent has a conical top with radius 2 m and a slant height of 2.5 m and a vertical height of 1.5 m. The lower part is a cylinder with radius 2 m and height 1 m.

(a) What is the total area of canvas?
(b) The tent sleeps 8 in comfort.
How much floor space is available per person?
(c) What is the total volume of the tent?

6 The figure shows a running track whose straight sides are *l* metres long. The ends are semi–circles each of diameter *d* metres. The perimeter, *P* metres of the track is given by the formula

$$P = 2l + \pi d.$$

(a) Make *l* the subject of this formula.
(b) If *P* = 400 m and *l* = 100 m calculate *d* (Take π = 3.14.)

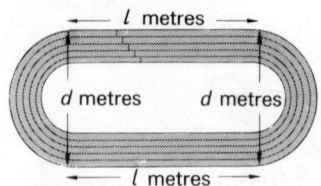

Tell your teacher you have finished this unit

UNIT 7 Mean, Median, and Mode

A Median

The boys are standing in order of size from the shortest to the tallest.

Pick out the one in the middle. He is called Fred.

Count the number who are shorter than Fred.

Count the number who are taller than Fred.

Here are the heights of the boys in centimetres:

148, 149, 150, (151,) 152, 153, 154

So Fred's height is 151 cm.

Exercise

1 The lengths of the boys' feet, measured in centimetres, are 22, 27, 24, 23, 25, 26, 23.

   (a) Write out a list of the sizes in order (smallest first).
   (b) Write down the size which is in the middle of your list.

2 A set of ages is given below:

14 years 9 months,	14 years 4 months,	15 years 1 month.
14 years 7 months,	15 years 2 months,	14 years 6 months.
15 years,	14 years 8 months,	14 years 5 months

   (a) Write them out in order (youngest first).
   (b) Write down the middle age.

> When a set of data is put into numerical order the measurement which occurs in the middle is called the **median** of the set.

Exercise

3 Write each set of numbers in order and then find its median:

   (a) 200, 195, 192, 205, 198   (b) 6, 7, 3, 4, 8, 5, 2, 12, 11
   (c) 1001, 880, 905, 870, 750, 1250, 600   (d) 22, 22, 17, 21, 20, 26, 38
   (e) 1, 3, 52, 61, 78, 82, 55, 76, 90, 94, 87, 65, 54

Example

The set 18, 14, 17, 27, 25, 20 can be put in order like this:

14, 17, (18, 20,) 25, 27

There is no middle number, so to find the median we take the two middle members, and go half way between them.

So median = 19

Exercise

4   Write each set of numbers in order and then find its median:

(a) 3, 11, 10, 15, 6, 8
(b) 4, 8, 9, 4, 5, 6
(c) 108, 92, 87, 140, 115, 98
(d) 23, 32, 38, 27, 30, 24
(e) 3.5, 4.5, 2, 6.5, 2.5, 3.5, 5, 5.5

Continue with Section B

B   Comparison using median

Do the boys look taller than the girls?
One way to compare them is to use the medians.

Example

The boys' heights in centimetres are:

151, 151, 152, 154, 155, 159, 159

and the girls' heights are:

148, 150, 152, 152, 154, 157, 159

The median height of the boys is 154 cm.
The median height of the girls is 152 cm.
The two groups can be roughly compared by comparing the medians.
The boys seem to be taller, since 154 is greater than 152.

Exercise

By putting the heights in order first, find the medians of the groups shown below and in each question state which group seems to be the taller.

1   Group A   162, 158, 160, 159, 157, 161, 158
Group B   159, 159, 163, 157, 156, 156, 157

2   Group A   120, 119, 124, 121, 118, 122, 120
Group B   125, 122, 122, 117, 116, 123, 115

Continue with Section C

C Median from frequency tables

The table opposite shows the number of mistakes made by pupils in a spelling test. How many pupils sat the test?

Number of mistakes	Frequency
0	2
1	3
2	7
3	8
4	5
5	4
6	1
Total	30

The table shows that
2 pupils had no mistakes
3 pupils had 1 mistake each
7 pupils had 2 mistakes each
and so on, so 30 pupils sat the test.

If we write out the number of mistakes made by each pupil we get:

0,0,1,1,1,2,2,2,2,2,2,2,3,3,3,3,3,3,3,3,4,4,4,4,4,5,5,5,5,6

We can now see that the median is 3.

Exercise

1 Find the median for each of the following sets of data.

(a)
Number of goals	Frequency
0	7
1	4
2	5
3	2
4	1
5	1
Total	20

(b)
Number of marks	Frequency
5	2
6	3
7	3
8	8
9	8
10	6
Total	30

(c)
Number of pets	Frequency
0	8
1	12
2	3
3	1
4	2
Total	26

Example

We will take the table at the top of the page and try to find the median without making a list.

Number of mistakes	Frequency
0	2
1	3
2	7
3	8
4	5
5	4
6	1
Total	30

The total frequency is 30.
This splits into 15 and 15, so the median lies between the 15th and 16th pupils.
2+3+7 = 12, so we have the first 12 pupils so far.
2+3+7+8 = 20, so we now have the first 20 pupils.
The 15th and 16th pupils must lie in the group of 8 pupils.
Each of these pupils had 3 mistakes, so the median is 3.

Exercise

2   Find the median for each of the following sets of data.

(a)

Score	Frequency
0	1
1	2
2	4
3	6
4	8
5	7
6	6
7	3
8	2
9	1
Total	40

(b)

Score	Frequency
10	1
11	2
12	4
13	5
14	6
15	3
16	3
17	1
Total	25

(c)

Score	Frequency
0	2
1	2
2	3
3	5
4	8
5	12
6	7
7	6
8	4
9	1
Total	50

Continue with Section D

D   Mean

Bill and Jack were at rifle practice.
Bill fired five shots and scored 8, 6, 9, 7, 8
Jack fired eight shots and scored 6, 9, 8, 9, 9, 8, 6, 5

If we arrange the scores in order we get

6, 7, 8, 8, 9   and   5, 6, 6, 8, 8, 9, 9, 9
    ↑   ↑
  median   median

Both sets have the same median, but there is another way of finding out who has the better performance.

$$\text{Bill's mean score} = \frac{38}{5} = 7.6$$

$$\text{Jack's mean score} = \frac{60}{8} = 7.5$$

```
 8      6
 6      9
 9      8
 7      9
 8      9
——      8
38      6
        5
       ——
       60
```

Bill has the higher mean, so his performance was better.

> To find the **mean** of a set of data remember that we add up all the data and divide by the number of items added.

Exercise

1   Find the mean (average) of each of the following sets:

(a) 2, 7, 3, 14, 9, 7, 13, 12, 8, 5   (b) 12, 23, 14, 17, 22, 19, 13, 20
(c) 4.7, 8.2, 7.3, 5.6, 4.9, 6.5, 7.6   (d) 184, 165, 171, 177, 180, 182
(e) 84   36   28   57   34   65   81   75   40   38
    67   72   77   35   29   31   88   51   58   33
    74   62   37   43   82   54   61   52   46   63
    57   60   36   78   87   69   75   84   36   45

2   The average of five numbers is 13. What is the total of the five numbers?

Example

The Prime Minister has changed her Cabinet. She has replaced the Chancellor of the Exchequer (who was aged 60) by another man (age 53) and the Minister of Education (who was aged 58) by a woman (age 45). The average age of the 20 members of the Cabinet used to be 57 years. What is the new average age?

Old Cabinet :   Average age of 20 members was 57 years
                  So the total of their ages was $20 \times 57 = 1140$ years
                  Total age of those leaving $= 60 + 58 = 118$ years
                  So total age of the remaining 18 members $= 1140 - 118 = 1022$ years

New Cabinet :   Total age of new members $= 53 + 45 = 98$ years
                  Total age of 20 members $= 1022 + 98 = 1120$ years
                  New average $= 1120 \div 20 = 56$ years

Exercise

3   The average age of sixty teachers at a school was 45 years. During the session four teachers aged 65, 60, 43, and 27 leave, and are replaced by four teachers aged 45, 40, 26, and 24.

Find the new average age of the sixty teachers.

4   The average weight of the Scottish rugby team which played Wales was 84 kg.

The team to play England has three changes. The players missed out had weights 81 kg, 87 kg, and 91 kg. The weights of the new players are 79 kg, 86 kg, and 88 kg.

What is the average weight of the new team of fifteen players?

Continue with Section E

E   Mean from frequency tables

Example

This table shows the scores of 30 pupils in a test question.
What is the mean score?

(It is easier to handle data when they are in a table like this, rather than listed as in question 1 of the last exercise.)

We find the total score like this :

Score	Frequency
0	5
1	6
2	8
3	7
4	3
5	1
Total	30

Score	Frequency	Score × frequency	
0	5	0	5 scores of 0, a total of 0
1	6	6	6 scores of 1, a total of 6
2	8	16	8 scores of 2, a total of $8 \times 2 = 16$
3	7	21	
4	3	12	
5	1	5	
Total	30	60	total score

$$\text{The mean score} = \frac{\text{total score}}{\text{number of pupils}} = \frac{60}{30} = 2$$

Exercise

1 In another question the scores were as shown in this table.

(a) Copy the table and calculate the mean score.

(b) State whether the pupils performed better on this question or on the one in the example above.

Score	Frequency	Score × frequency
0	6	
1	7	
2	7	
3	5	
4	4	
5	1	
Total	30	

2 Copy each of these tables and calculate the mean:

(a)

Score	Frequency
10	3
15	4
20	8
25	4
30	1
Total	

(b)

Score	Frequency
0	4
1	5
2	7
3	9
4	2
5	3
Total	

(c)

Score	Frequency
0	2
4	5
8	6
12	4
16	3
Total	

(d)

Score	Frequency
0	1
5	3
10	4
15	4
20	6
25	8
30	4
Total	

Continue with Section F

F Mean from bar chart

The bar chart shows the total number of goals scored in 60 hockey matches. Calculate the mean.

To calculate the mean we complete a frequency table like those shown in Section E.

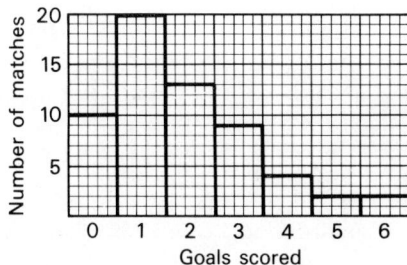

Goals scored	Frequency	Score × frequency
0	10	0
1	20	20
2	13	26
3	9	27
4	4	16
5	2	10
6	2	12
Totals	60	111

$$\text{Mean} = \frac{111}{60} = 1.85$$

Exercise

1 The bar chart shows the number of passengers carried by a minibus in 50 journeys.

Calculate the mean number of passengers per journey.

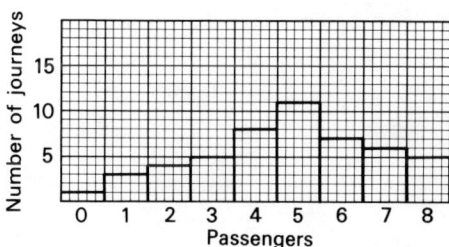

2 This bar chart shows the number of accidents in which a group of 80 bus drivers were involved.

Calculate the mean number of accidents per driver.

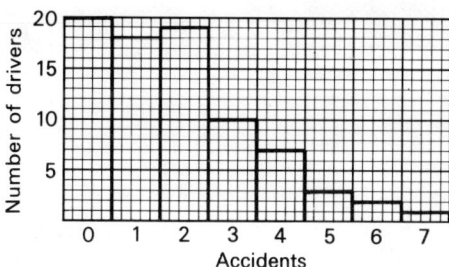

Continue with Section G

G   Mode

The manager of a small shoe shop is ordering a new style of girls' shoe. He checks his records to see how many of each size he should order.

When the shoes arrive the boxes are stacked as shown here.

He is not interested in the mean or the median of this set.

When ordering, however, he has to decide which size will be most in demand. From the diagram we see he expects to sell more of size 4 than of any other size.

For this set we say the **mode** is 4.

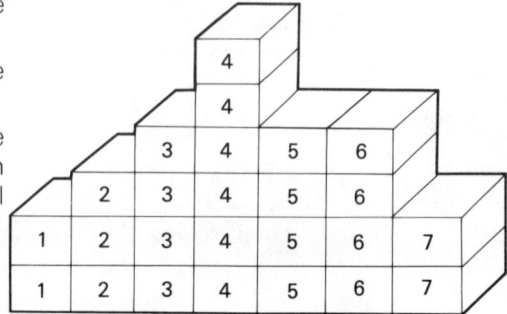

The **mode** of a set is the number which occurs most frequently.

Exercise

Shirts by Lees	To: W. R. Allan
Size	Number of shirts
14	x x x
14½	x x x x x . x
15	x x x x
15½	x x x
16	x x

1 A shopkeeper has just received and checked a parcel containing men's shirts. The invoice shows the sizes supplied and the number of each size supplied. What is the mode?

2 For each of the following sets of data, state the mode.

(a)

Number of pets	Frequency
0	8
1	12
2	3
3	1
4	2
Total	26

(b)

Number of goals	Frequency
0	7
1	4
2	5
3	2
4	1
5	1
Total	20

(c)

Score	Frequency
0	1
5	3
10	4
15	4
20	6
25	8
30	4
Total	30

3 In a traffic survey the number of persons in each car which passed was recorded. Here are the data.

```
1  2  1  4  2  3  2  1  1  4  2  1  3  1  2  2  1  1
2  1  4  2  4  3  1  2  1  1  2  1  3  1  4  4  1  1
1  2  1  2  2  1  3  1  2  3  1  2  1  1  2  3  1  2
3  1  4  2  1  2  3  2  1  1  2  4  2  3  1  2  2  1
4  2  1  2  2  1  1  3  2  1  3  2  1  4  2  1  1  2
```

(a) Using tally marks, construct a frequency table for the above data.
(b) Write down the mode.

Continue with Section H

H Range

Example

What is the range of the following set of numbers?

$$11, 4, 8, 3, 7, 12, 6, 16, 10$$

The lowest is 3 and the highest is 16.
So the range is 3 to 16.
This is a range of 13.

$$\begin{array}{r} 16 \\ -\ 3 \\ \hline 13 \end{array}$$

Exercise

For each of the following sets write down (a) the lowest number, (b) the highest, and (c) the range:

1 7, 9, 10, 14, 16, 19, 21, 28, 32

2 8, 7, 17, 6, 4, 5, 11, 9, 5

3 27, 24, 28, 23, 21, 26, 29, 22, 25

4 14.5, 13, 14.3, 12.6, 13.4

5
84	36	28	57	34	65	81	75	49	38	67	72	77	35
29	31	88	51	58	33	74	62	37	43	82	54	61	52
46	63	57	60	36	78	87	69	75	84	36	45	50	66

Continue with Section I

I Summary

We have used three different types of representative of a set.
Of the three, the most commonly used is the mean.

$$\text{The } \mathbf{mean} = \frac{\text{the total of all the data}}{\text{the number of data items}}$$

It is sometimes more convenient to use the median, if it will involve less effort to find than the mean.

$$\text{The } \mathbf{median} = \text{the middle value of the data.}$$

In some situations neither the mean nor the median is appropriate, and the mode is used.

$$\text{The } \mathbf{mode} = \text{the most frequently occurring of the data.}$$

Exercise

1 This bar chart shows a symmetrical distribution.

   (a) Write down the mode.
   (b) Make a frequency table for the data in the bar chart.
   (c) Find the median.
   (d) Calculate the mean.
   (e) Copy and complete:

      When the data are symmetrical, the mean, median, and mode are _____.

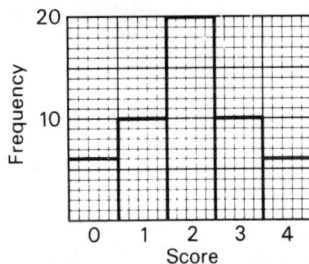

Continue with Section J

J Progress check

Exercise

1 Find the median of each of the following sets:

    Set A   26, 5, 14, 22, 18, 6, 9, 22, 17, 19, 29
    Set B   24, 26, 20, 28, 22, 25, 26, 22, 28

    State the range of each set.

2 For the data shown in this bar chart,

    (a) state the mode.
    (b) make a frequency table.
    (c) find the median.

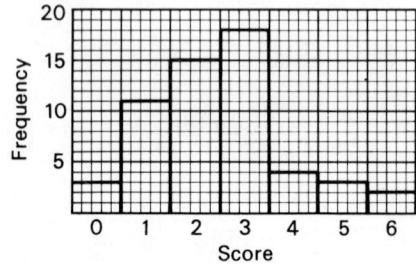

3

Mark	Frequency
5	1
6	3
7	4
8	7
9	8
10	3
11	2
12	2

The table shows the marks of thirty pupils in an English test.

Calculate the mean mark.

Ask your teacher what to do next

K Median class

A class of 32 pupils sat an English test. The marks are given below.

32	84	76	63	30	51	42	38	67	64	35	40	55	68	46	59
57	36	48	29	56	54	60	47	59	73	70	41	40	58	47	48

The range is from 29 to 84, a range of 55.
This is so large that to make a frequency table for these data we must use class intervals like this:

Mark	Tally	Frequency
20–29	I	1
30–39	ЖТ	5
40–49	ЖТ IIII	9
50–59	ЖТ III	8
60–69	ЖТ	5
70–79	III	3
80–89	I	1
	Total	32

The total frequency is 32.
This splits into 16 and 16. So the median lies between the 16th and 17th pupils.
1 + 5 + 9 = 15, so the 16th and 17th pupils are in this group.

The median class is 50–59.

Exercise

1 (a) Using the same class intervals as above, make a frequency table for the following data:

52	23	46	37	64	51	73	82	65	27	42	39	60
47	60	58	36	43	75	66	64	57	68	42	77	61

(b) What is the range of the data above?
(c) Find the median class.

2 The weights of calves are recorded when they are sold. The bar chart gives the weights for one lot.

(a) Make a frequency table for these data.
(b) Find the median class.
(c) State the maximum range of the weights.

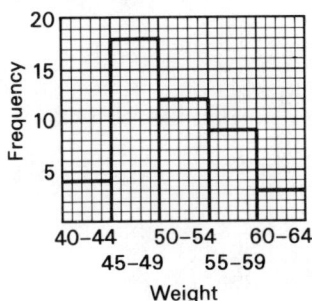

3 In an experiment to check on the progress of trees, leaf areas were measured in square centimetres. The following data were obtained.

47	72	53	64	58	79	62	68	55	45	66	70
64	75	59	48	63	42	50	78	44	67	59	60
46	79	68	72	57	58	64	58	61	55	59	62

(a) Find the range of the data.
(b) Using class intervals like 50–54, 55–59, make a frequency table for the data.
(c) Find the median class.

4 For the data shown in this bar chart,

(a) make a frequency table.
(b) find the median class.
(c) state the maximum range of the data.

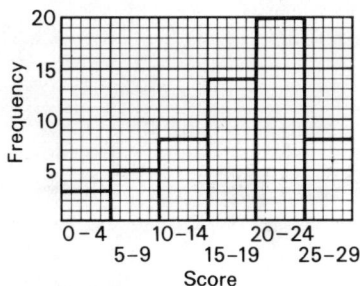

Continue with Section L

L Mean from grouped data

The time spent by 50 machine operators on a piece of work was recorded and is shown in this table using class intervals.

Time in seconds	Frequency
60–64	4
65–69	8
70–74	10
75–79	12
80–84	9
85–89	7
Total	50

Without knowing the time taken by each operator we cannot find the mean exactly.

We can, however, make a fairly accurate estimate by assuming that in the group who took from 60 to 64 seconds the average time taken was at the mid–point of 60, 61, **62**, 63, 64.

> In each case we assume that the average time taken by the group is the mid–point of the class interval for that group

For those taking from 65 to 69 seconds we take the average time as 67.
For those taking from 70 to 74 seconds we take the average time as 72, and so on.

An estimate of the total time taken by the groups is now found.

Time in seconds	Frequency	Mid–point of class interval	Frequency × mid–point
60–64	4	62	$4 \times 62 = 248$
65–69	8	67	$8 \times 67 = 536$
70–74	10	72	$10 \times 72 = 720$
75–79	12	77	$12 \times 77 = 924$
80–84	9	82	$9 \times 82 = 738$
85–89	7	87	$7 \times 87 = 609$
Totals	50		3775

An estimate of the mean is therefore $\dfrac{3775}{50} = 75.5$ seconds

Exercise

1   (a) Copy and complete this table showing the marks scored in a test by 30 students.

Marks	Frequency	Mid–point of class interval	Frequency × mid–point
30–34	2	32	64
35–39	5	37	185
40–44	10	42	
45–49	5		
50–54	5		
55–59	3		
Totals	30		

(b) Find the mean by dividing the total of the end column by the total of the frequency column.

2 Using the method of question 1 calculate the mean for the following data.

(a)

Height (cm)	Frequency	Mid–point	Frequency × mid–point
148–150	1	149	
151–153	2		
154–156	4		
157–159	10		
160–162	2		
163–165	1		
Totals	20		

$$\text{Mean} = \frac{\quad}{20} =$$

(b)

Time in seconds	Frequency
10–12	2
13–15	6
16–18	7
19–21	7
22–24	5
25–27	3
Total	30

(c)

Score	Frequency
0– 4	4
5– 9	6
10–14	20
15–19	10
20–24	5
25–29	5
Total	50

3 Copy and complete this table, and calculate the mean.

Mass (g)	Frequency	Mid–point	Frequency × mid–point
0– 5	2	2.5	5
6–11	4	8.5	34
12–17	4	14.5	
18–23	6	20.5	
24–29	10		
30–35	4		
Totals	30		

4 Calculate the mean from these data.

Score	Frequency
0– 9	10
10–19	15
20–29	20
30–39	10
40–49	5
Total	60

Continue with Section M

M Interpretation of data

Example

The bar chart shows the accident record of drivers at a bus depot, over a period of one year.

(a) How many drivers were there?
Number of drivers $= 27+5+3+1 = 36$
(b) How many drivers had accidents?
27 drivers had no accidents, so number having accidents $= 5+3+1 = 9$
(c) What percentage of drivers had accidents?
Percentage $= \frac{9}{36} = \frac{1}{4} = 25\%$
(d) The most appropriate representative of this set is the mode. What is the mode? The mode is 0 accidents.

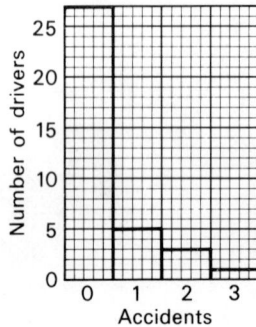

(e) If a driver is picked at random what is the probability that he had two accidents?
3 of the 36 drivers had two accidents, so probability $= \frac{3}{36} = \frac{1}{12}$.
(f) If a driver is picked at random what is the probability that he had two or more accidents?
'two or more' means in this case two accidents or three accidents. 4 of the 36 drivers had two or more accidents, so probability $= \frac{4}{36} = \frac{1}{9}$.

Exercise

1   The bar chart shows the membership of a teenage youth club according to age.

   (a) Make a frequency table for these data.
   (b) How many members are there?
   (c) Calculate the mean age.
   (d) What percentage of the membership is older than 15?
   (e) If a member is picked at random what is the probability of the member's age being (i) 15, (ii) less than 15?

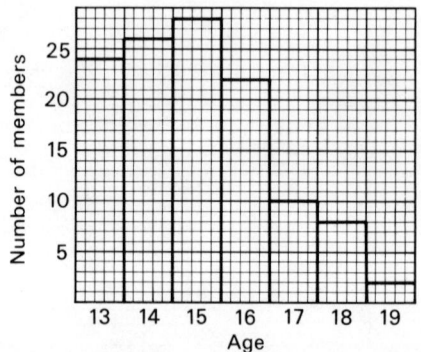

2   A survey of the amount of time spent doing homework during one week was made within a class.

   The times were recorded in hours to the nearest half hour.

6	7	$5\frac{1}{2}$	5	$7\frac{1}{2}$	5	6	$4\frac{1}{2}$	5	$4\frac{1}{2}$
6	$7\frac{1}{2}$	$4\frac{1}{2}$	6	8	5	$7\frac{1}{2}$	6	$5\frac{1}{2}$	6
7	$6\frac{1}{2}$	6	$6\frac{1}{2}$	7	$5\frac{1}{2}$	$6\frac{1}{2}$	$5\frac{1}{2}$	6	$5\frac{1}{2}$

   (a) Draw a frequency table and record these data in it.
   (b) Calculate the mean time in hours.
   (c) What percentage of pupils in the class spent less than the mean time on their homework?
   (d) If a pupil is picked at random what is the probability that the pupil spent $6\frac{1}{2}$ hours or more on homework?

3 This frequency table gives the marks obtained by a group of pupils in an English examination.

Mark	Frequency
0– 9	0
10–19	2
20–29	8
30–39	9
40–49	14
50–59	24
60–69	23
70–79	21
80–89	15
90–99	4

(a) How many pupils were in the group?

(b) If the pass mark is set to allow 50% of the pupils to pass, in which class interval must the pass mark lie?

(c) If instead the pass mark is to be set to allow 75% of the pupils to pass, in which class interval must the pass mark lie?

(d) Which is the modal class?

(e) What is the maximum range of the marks gained by the pupils?

4 The bar chart shows the expected distribution of scores when two dice are thrown 36 times, and the total score recorded each time.

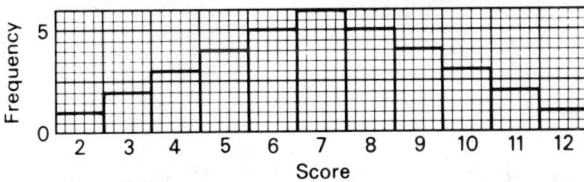

Use the bar chart to answer the following questions.

(a) What is the probability of getting a total score of 7 with two dice?

(b) What is the probability of getting a total score of 10 or more with two dice?

(c) What is the modal score?

(d) What percentage of scores will be more than 4 but less than 7?

5 Four dice are thrown and the total score is recorded.

(a) What is the lowest possible score?

(b) What is the highest possible score?

(c) Which of the following class intervals are not needed when recording the scores?

   1–3, 4–6, 7–9, 10–12, 13–15, 16–18, 19–21, 22–24, 25–27, 28–30

(d) You are going to throw four dice forty times.

   Draw a frequency table using class intervals like those above, and with a tally column, and then record the results of your throws.

(e) From your table, what is the probability of getting a score of from 4 to 6?

(f) Which is the modal class?

(g) What percentage of scores lie in the modal class?

(h) Calculate the mean score.

Continue with Section N

N Progress check

Exercise

1   Write down the mid–point of each of the following class intervals:

(a) 70–74   (b) 10–19   (c) 58–62

2   Calculate the mean of the following set of data.

Number of seeds germinating	Frequency
0– 4	1
5– 9	5
10–14	10
15–19	6
20–24	5
25–29	3
Total	30

3   This bar chart shows the shoe sizes of a group of boys.

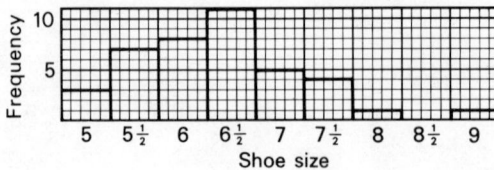

Shoe size

(a) What is the modal size?
(b) What percentage of boys take a shoe size less than the mode?
(c) If a boy is picked at random from this group, what is the probability that his shoe size will lie in the range 6 to 7?

4   The table gives the distribution of the ages of the people in a village.

(a) State the modal class.
(b) Find the median class.
(c) What is the maximum range of the ages?
(d) What percentage of the people in the village are under 20 years old?
(e) A teenager is one whose age is from thirteen to nineteen. What is the maximum number of teenagers?

Class interval	Number of people
0– 9	28
10–19	34
20–29	25
30–39	24
40–49	30
50–59	25
60–69	14
70–79	10
80–89	6
90–99	4
Total	200

Tell your teacher you have finished this unit

UNIT 8 Borrowing and Saving

A Bank accounts

A Savings Bank is a safe and convenient place to keep your money. The customer can deposit money or withdraw money by presenting his passbook at the bank. Many employers pay wages and salaries directly into their employees' bank accounts by what is called **credit transfer.**

Telephone, electricity, and gas bills can be paid easily at a bank by withdrawing money from the account.

Recurring payments such as hire–purchase instalments, mortgage repayments, and insurance premiums can be made automatically by the bank. If the customer has signed a **standing order** to make one of these payments, the bank will continue to make regular payments, possibly each month, until the order is cancelled.

Here is a page from a passbook:

BURGH SAVINGS BANK *in Account with* ANDREW MURRAY

A/c. No. 34712/6

Date			Deposits		Withdrawals		Balance due	
		Brought forward					34	72
Oct	1	Salary (CT)	580	50			615	22
	2	Self (cash)			100	00	515	22
	15	Building Society (SO)			152	12	363	10
	17	Electricity Bill			91	81	271	29
	22	Income tax rebate	39	05			310	34
	23	Insurance Premium (SO)			43	27	267	07
Nov	1	Salary (CT)	581	70			848	77

Explanation: £34.72 was brought forward from the previous page.

Oct. 1 — Mr Murray's salary was paid into the bank account by credit transfer (CT).

Oct. 2 — Mr Murray withdrew £100 in cash.

Oct. 15 — The bank made a regular payment to the Building Society according to a standing order (SO).

Oct. 17 — Mr Murray visited the bank to pay an electricity bill.

Oct. 22 — Mr Murray paid in a cheque for £39.05.

Oct. 23 — The bank made a regular payment for an insurance premium.

Exercise

Copy and complete the following passbook pages using the information given:

1 July 1 – Monthly salary of £610.50 paid into account.
   July 2 – £80 withdrawn in cash.
   July 8 – Monthly Building Society payment of £146.71 was made according to standing order.
   July 12 – Monthly hire-purchase instalment of £37.25 was made according to standing order.
   July 22 – Share dividend of £43.10 paid into account.
   July 29 – Gas bill of £53.46 paid.

BURGH SAVINGS BANK *in Account with*   JOHN BROWN

A/c. No. 97215/1

Date			Deposits		Withdrawals		Balance due	
		Brought forward					138	72
July	1							
	2							
	8							
	12							
	22							
	29							

2 December 2 – Monthly salary of £487.40 paid into account.
   December 4 – £60 in cash withdrawn.
   December 11 – £33.73 paid for telephone bill.
   December 20 – Postal order for £10 paid into account.
   December 23 – Annual insurance premiums of £47.20 paid by standing order.
   December 31 – Annual interest of £27.45 credited to account.

BURGH SAVINGS BANK *in Account with*   JAMES BLACK

A/c. No. 20781/6

Date			Deposits		Withdrawals		Balance due	
		Brought forward					272	41
Nov	28	Corporation rates			109	70	162	71
Dec	2							

Continue with Section B

B Simple interest

Money which is deposited in a Bank or Building Society is known as the **principal** or **capital**.

The money which the bank pays for the use of the principal is called **interest**.

Sometimes people talk about **investing** money to receive interest.

Exercise

A man put £630 in a Savings Bank which offers an interest rate of $7\frac{1}{2}\%$ per year. What interest will the man receive at the end of a year?

$$\text{Interest} = 7\frac{1}{2}\% \text{ of } £630$$
$$= £47.25$$

Working

$$1\% \text{ of } £630 = £ 6.30$$
$$\times 7$$
$$7\% \text{ of } £630 = £44.10$$
$$\frac{1}{2}\% \text{ of } £630 = £ 3.15$$
$$7\frac{1}{2}\% \text{ of } £630 = £47.25$$

Divide by 2.

Exercise

1 The Trustee Savings Bank offers 8% per annum in their Special Investment Department. ('per annum' means 'for a year') Find the interest on £315 for a year.

2 To pay for a new machine a company borrows £6000, the rate of interest being $8\frac{1}{2}\%$. What interest is due at the end of the first year?

3 Find the interest payable for one year on the following principals invested at the rates given:

	Principal	Rate of interest
(a)	£600	4%
(b)	£340	$4\frac{1}{2}\%$
(c)	£40	$7\frac{1}{4}\%$
(d)	£800	$7\frac{1}{2}\%$
(e)	£640	$7\frac{1}{4}\%$

4 To buy new machinery a firm borrows £8000 for a year from a bank, the interest rate being 9%. What sum (that is, capital and interest) must be repaid at the end of the year?

Example

WINDMILL BUILDING

WINDMILL BUILDING SOCIETY 9½%

A man invested £260 in a Building Society whose rate of interest is $9\frac{1}{2}\%$ per annum. How much interest would he receive at the end of the first six months?

Working

Interest for 1 year $= 9\frac{1}{2}\%$ of £260
$\qquad\qquad\qquad\quad = £24.70$

1% of £260 $= £\ 2.60$	
$\times 9$	Divide
9% of £260 $= £23.40$	by 2.
$\frac{1}{2}\%$ of £260 $= £\ 1.30$	
$9\frac{1}{2}\%$ of £260 $= £24.70$	

Interest for 6 months $= \frac{1}{2}$ of £24.70
$\qquad\qquad\qquad\qquad\ = £12.35$

Exercise

5 A man borrowed £360 from a Finance Company at the rate of 12% per annum. How much interest would he pay after 3 months?

6 If the rate for Building Society investors is $9\frac{1}{2}\%$. What half–yearly interest would you receive on a deposit of £820?

7 Find the interest payable on the following principals invested at the rates and for the period of time given:

	Principal	Rate of interest	Period of time
(a)	£160	7%	1 year
(b)	£500	10%	6 months
(c)	£420	3%	3 months
(d)	£3000	$9\frac{1}{2}\%$	4 months
(e)	£480	$6\frac{1}{2}\%$	1 month

8 A man invests £440 in $8\frac{1}{2}\%$ British Savings Bonds. If he decides to cash the Bonds after 6 months, how much money should he receive back?

Continue with Section C

C Compound interest

In his Will, General I. M. Goodsir left £400 in the care of Trustees. He requested that the interest from the investment of this capital be given each year to the eldest retired officer living in the city.

From 1981 to 1983, the interest from the Goodsir Bequest was claimed by ex-Major D. K. Wilnot.

For these three years the interest paid at 5% per annum was as follows:

		Working		
1981 Interest	= £20.00	1% of £400 =	£ 4	
1982 Interest	= £20.00		× 5	
1983 Interest	= £20.00			
1981–1983 Total Interest	= £60.00	5% of £400 =	£20	

Since Major Wilnot's death, the Bequest has remained unclaimed and the interest has been accumulating in the Bank.

Here are the accounts for the three years which followed:

Working

1984	Capital	= £400.00	1% of £400 = £ 4.00
	Interest	= £ 20.00	× 5
	Amount	= £420.00	5% of £400 = £20.00
1985	Capital	= £420.00	1% of £420 = £ 4.20
	Interest	= £ 21.00	× 5
	Amount	= £441.00	5% of £420 = £21.00
1986	Capital	= £441.00	1% of £441 = £ 4.41
	Interest	= £ 22.05	× 5
	Amount	= £463.05	5% of £441 = £22.05
	Final Amount	= £463.05	
	Original Capital	= £400.00	
1984–1986	Total Interest	= £ 63.05	

The interest which has accumulated in this way is called **compound interest**.

Notice that the interest paid out in the years 1981, 1982, and 1983 is £60 but the compound interest for the next three years is *more than* £60.

Exercise

1   Calculate the compound interest on investing £500 for 2 years at 8% per annum.

Copy and complete:

Working

1st year	Principal	= £500.00	1% of £500 = £ 5
	Interest	= £ 40.00◄	× 8
	Amount	= £540.00	8% of £500 = £40
2nd year	Principal	£540.00	1% of £540 = £ 5.40
	Interest	£▨▨▨▨◄	× 8
	Amount	£▨▨▨▨	8% of £ £▨▨▨

Final Amount   = £▨▨▨▨
Original Principal = £▨▨▨▨
Compound Interest = £▨▨▨▨

2   Calculate the compound interest on investing £800 for 3 years at 5% per annum.

Copy and complete:

Working

1st year	Principal	= £800.00	1% of £800 = £ 8
	Interest	= £ 40.00◄	× 5
	Amount	= £840.00	5% of £800 = £40
2nd year	Principal	= £840.00	1% of £▨ =
	Interest	= £▨▨▨◄	× 5
	Amount	= £▨▨▨	5% of £▨ = ▨▨▨
3rd year	Principal	= £▨▨▨	1% of £▨ = ▨▨▨
	Interest	= £▨▨▨◄	× ▨
	Amount	= £▨▨▨	= ▨▨▨

Final Amount   = £▨▨▨
Original Principal = £▨▨▨
Compound Interest = £▨▨▨

3   Calculate the compound interest on investing:

(a) £600 for 2 years at 5% per annum.
(b) £800 for 3 years at 10% per annum.

4   Calculate the compound interest on a deposit of £1000 for 3 years if the rate for the first year is 5%, the second year 6%, and the third 7%.

5   Calculate the compound interest on £6000 invested at 10% for 4 years.

Continue with Section D

D Building societies

YORKSHIRE
ABBEY
NATIONAL
BUILDING SOCIETY
HALIFAX
WOOLWICH
Nationwide

Money invested in a Building Society is loaned to people to enable them to buy their own houses. The people who invest money (the investors) receive interest and the people who are loaned money (the borrowers) pay interest.

Interest is paid to investors half–yearly (in some Building Societies in June and December). The investor can either receive a cheque for the interest *or* leave the interest in his Account to accumulate more interest.

Example

A man invested £820 in Safeway Building Society at the beginning of 1985. If the rate of interest remained steady at 7% and is added to the account half-yearly, calculate the compound interest which has accumulated after 1 year.

Working

1985 June   Principal = £820.00   1% of £820 = £ 8.20
           Interest = £ 28.70   × 7
           Amount = £848.70   7% of £820 = £57.40
                           For half-year = £28.70

             | Interest is calculated on whole £s only |

December   Principal = £848.70   1% of £848 = £ 8.48
           Interest = £ 29.68   × 7
           Amount = £878.38   7% of £848 = £59.36
                           For half-year = £29.68

       Final Amount = £878.38
   Original Principal = £820.00
  Compound Interest = £ 58.38

Exercise

1   Calculate the Building Society interest on investing £450 for 1 year at 9% compounded half–yearly.

Copy and complete:

Working

1st year	June	Principal = £450.00	1% of £450 = £ 4.50
		Interest = £ 20.25 ◄	× 9
		Amount = £470.25	9% of £450 = £40.50
			For half–year = £20.25

Calculate interest on whole £s only.

	December	Principal = £470.25	1% of £470 = £▓
		Interest = £▓	× 9
		Amount = £▓	9% of £ = £▓
			For half–year = £▓

Final Amount   = £▓
Original Principal = £▓
Compound Interest = £▓

2   Calculate the Building Society Interest on investing £280 for 2 years at 8% compounded half–yearly.

Copy and complete:

Working

1st year	June	Principal = £280.00	1% of £280 = £ 2.80
		Interest = £ 11.20 ◄	× 8
		Amount = £291.20	8% of 280 = £22.40
			For half–year = £11.20

Calculate interest on whole £s only.

	December	Principal = £291.20	1% of £291 = £▓
		Interest = £▓	
		Amount = £▓	

2nd year	June	Principal = £▓	
		Interest = £▓	
		Amount = £▓	

	December	Principal = £▓	
		Interest = £▓	
		Amount = £▓	

Final Amount   = £▓
Original Principal = £▓
Compound Interest = £▓

Continue with Section E

E   SAVE

A Building Society Scheme
for
Regular Savers

Under the Save-As-You-Earn scheme if you are prepared to enter into a contract to save regularly every month for five years with the Society you will be entitled, at the end of that period, to a bonus equivalent to one year's savings and on this bonus neither you nor the Society will be liable for any form of taxation.

The bonus will be equivalent to a compound rate of interest of about 7 per cent per annum.

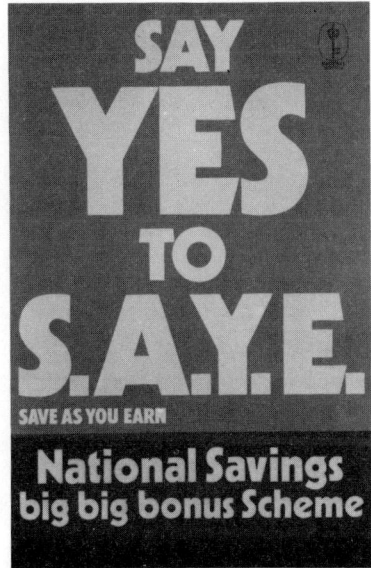

To avoid difficult compound interest calculations, the Building Societies provide ready reckoners.

Here are examples of how your savings will accumulate

Amount saved each month	Total savings after 5-years	Savings plus 5-year bonus
£	£	£
1	60	72
2	120	144
3	180	216
4	240	288
5	300	360
6	360	432
7	420	504
8	480	576
9	540	648
10	600	720

Example

What compound interest would you receive under SAYE if you saved £5 per month?

Savings plus 5-year bonus = £360
Total savings after 5 years = £300
Compound Interest   = £ 60

Exercise

Using the table above, find the compound interest you would receive under SAYE at the end of 5 years if you saved:

1   £3 per month   **2**   £15 per month   **3**   £17 per month

Continue with Section F

F Buying a house

We have discussed how a Building Society receives money from investors. Now let us consider the position of people who wish to borrow money to buy their own house.

The buyer is said to be purchasing the house on a mortgage. He pays back the money to the Building Society, plus interest, over a long period, sometimes as long as 30 years.

Valuation

Before giving a loan, the Society send their valuer to inspect the property and estimate its value. The scale of fees is as follows:

Valuation	Fee*
Up to £15 000	£31
Over £15 000	£31 plus £1 for each £1000 or part of £1000 in excess of £15 000

* The valuation fees do not include value added tax (VAT).

Example

What is the bill for the valuation of a house valued at £23 000. (Take the VAT rate as 15%.)

Value of house = £23 000
Basic fee = £31

£23 000 is £8000 more than £15 000 and £8000 = 8 × £1000.

So additional fee = £8
Total fee = £31 + £8 = £39
VAT = 15% of £39 = £5.85
Total bill = £39 + £5.85 = £44.85

Exercise

Calculate the total bill for each of the following valuations.
(Take the VAT rate as 15%.)

1 £26 000	2 £20 000	3 £14 000	4 £33 400

5 £39 000

6 £56 500

Continue with Section G

G House loans

Building Societies normally advance a loan which usually amounts to between 80% and 95% of the **value** of the house. On new houses, some Societies provide 100% mortgages.

Building Societies provide loans based on their own valuation, not on the buying price.

Example

A man agrees to pay £26 000 to buy a house which the Building Society values at £24 000. If the Society have offered to advance an 80% loan, (a) how much does the society lend, (b) what deposit will the buyer need to provide?

Working

(a) Loan = 80% of the valuation
     = £19 200

1% of £24 000 = £ 240
                 × 80
80% of £12 000 = £19 200

(b) Deposit = Buying Price − Loan
     = £26 000 − £19 200
     = £6800

Exercise

For each of the following house purchases calculate (a) the amount of the loan and (b) the deposit the buyer must provide. Set out your working as above.

	Valuation	Buying price	Loan %
1	£26 000	£27 000	80%
2	£20 000	£20 500	85%
3	£18 600	£19 000	90%
4	£31 000	£32 000	100%

Continue with Section H

H Loan repayment

The normal method of repaying the loan and interest is by making fixed monthly instalments. The amount of the instalment depends on three factors:

1   The amount of the loan.
2   The number of years over which the loan is to be repaid.
3   The current rate of interest being charged.
4   The current rate of income tax relief.

The following table shows the monthly repayment for each £1000 of the loan to be repaid:

Mortgage Rates

Monthly Repayment on loan of £1000	Term of years			
	10	15	20	25
	£	£	£	£
11%	12.25	9.55	8.29	7.60

Example

A loan of £19 600 is made on a house for a period of 25 years at 11%.

(a) What is the amount of the monthly repayment?
(b) How much does the borrower pay over the term of the loan?
(c) How much interest will be paid over this term?

(a) Monthly Repayment $= £19\,600$ at £7.60 per £1000 $= 19.6 \times £7.60$
$= £148.96$

(b) Total Repayments over 1 year $= 12 \times £148.96$ $= £1787.52$
Total Repayments over 25 years $= 25 \times £1787.52$ $= £44\,688$

(c) Interest paid $=$ Total payments $-$ Loan $= £44\,688 - £19\,600$
$= £25\,088$

Exercise

For each of the following loans calculate (a) the monthly repayment, (b) the total repayments over the term of the loan, and (c) interest paid over this term.

	Loan	Term of years
1	£15 000	20
2	£18 000	25
3	£20 000	15
4	£22 500	25

Continue with Section I

House insurance

Fire Insurance: to insure his house and furniture against fire damage, a house-holder enters into a contract with an insurance company.

In this contract, called a **policy**, the company promises to pay back the value of the property and contents if damaged or destroyed by fire.

For his part, the householder pays the insurance company an annual payment called a **premium**.

Comprehensive Insurance: for a larger premium householders can be insured against fire, theft, storm damage, and so on. Such an insurance policy is called a **comprehensive insurance** cover.

Here are the rates charged by one insurance company:

Annual Premium per £100			
Fire Insurance		Comprehensive Insurance	
Building	*Contents*	*Building*	*Contents*
16p	24p	30p	60p

Minimum premium for each policy is £10.00.

Example

Calculate the total annual premium payable on a building valued at £24 000 and contents of £4000 to be covered comprehensively.

Building's
premium = £24 000 at 30p per £100
= 240 × 30p
= 7200p
= £72.00

Content's
premium = £4000 at 60p per £100
= 40 × 60p
= 2400p
= £24.00

Total premium = £72.00 + £24.00
= £96.00

Exercise

Calculate the total annual premium payable on each of the following:

	Cover	*Value of* Buildings	Contents
1	Comprehensive	£28 000	£3000
2	Fire	—	£2500
3	Fire	£18 000	£2000
4	Comprehensive	£36 000	£5000

Continue with Section J

J Life assurance

There are two main types of assurance policies – an **endowment policy** and a **whole-life policy.**

Through an **endowment policy** a person can assure himself of a fixed sum of money at the end of the term of the policy. If the assured person dies at any time before the policy has matured, his dependants receive the fixed sum assured.

In the case of **whole-life** assurance, the policy is only payable on the death of the assured person.

LIFE ASSURANCE WITHOUT PROFITS

MONTHLY PREMIUMS PER £1000 SUM ASSURED

For all sums assured add a Fixed Charge of £0.30 per policy

| ENDOWMENT ASSURANCE | | | | | | | WHOLE LIFE |
| payable at the end of the term or on previous death | | | | | | | payable at death |

Age Next Birthday	TERM							Age Next Birthday	
	15 years	20 years	25 years	30 years	To Age 55	To Age 60	To Age 65		
	£	£	£	£	£	£	£		£
20	4.35	2.98	2.17	1.65	1.30	1.05	0.88	20	0.60
21	4.35	2.98	2.17	1.66	1.37	1.10	0.92	21	0.62
22	4.35	2.98	2.17	1.66	1.43	1.15	0.96	22	0.64
23	4.36	2.98	2.18	1.66	1.51	1.21	1.00	23	0.67
24	4.36	2.98	2.18	1.67	1.58	1.26	1.04	24	0.69
25	4.36	2.98	2.18	1.67	1.67	1.33	1.09	25	0.72
26	4.36	2.98	2.18	1.67	1.76	1.39	1.14	26	0.75
27	4.36	2.99	2.19	1.68	1.86	1.46	1.19	27	0.78
28	4.36	2.99	2.19	1.69	1.96	1.54	1.25	28	0.81
29	4.36	2.99	2.20	1.69	2.08	1.62	1.31	29	0.84
30	4.36	2.99	2.20	1.70	2.20	1.70	1.37	30	0.87
31	4.37	3.00	2.21	1.71	2.34	1.80	1.44	31	0.91
32	4.37	3.00	2.22	1.72	2.49	1.90	1.51	32	0.95
33	4.37	3.01	2.23	1.74	2.65	2.00	1.59	33	0.99
34	4.38	3.02	2.24	1.75	2.83	2.12	1.68	34	1.03
35	4.38	3.03	2.25	1.77	3.03	2.25	1.77	35	1.08
36	4.39	3.04	2.26	1.78	3.25	2.39	1.86	36	1.13

Example

A man aged 26 next birthday wishes to take out an endowment policy for £13 500, payments being made over a 25-year period. Calculate his monthly premiums from the above tables.

Basic premium = £13 500 at £2.18 per £1000 = 13.5 × £2.18 = £29.43

Monthly premium = £29.43 + £0.30 (fixed charge) = £29.73

Exercise

1 A lady aged 32 next birthday takes out a whole life policy for £20 000.
   Calculate her monthly premium.
2 Find the monthly premium for an endowment policy of £12 500, payments being made
   over a 30–year period by a man aged 27 years at his next birthday.
3 Find the monthly premium for an endowment policy for £14 000, payments being made up
   to the age of 60 years by a man who will be 35 years old next birthday.

With profits

Policies are obtained 'with profits' or 'without profits'. A 'with profits' policy allows
the assured person a share in the Company's profits for which he has to pay a slightly
increased premium. This helps to keep the sum assured in step with inflation.

LIFE ASSURANCE WITH PROFITS
MONTHLY PREMIUMS PER £1000 SUM ASSURED

For all sums assured add a Fixed Charge of £0.30 per policy

ENDOWMENT ASSURANCE — payable at the end of the term or on previous death

WHOLE LIFE — payable at death

Age Next Birthday	TERM							Age Next Birthday	
	15 years	20 years	25 years	30 years	To Age 55	To Age 60	To Age 65		
	£	£	£	£	£	£	£		£
21	6.06	4.61	3.73	3.14	2.79	2.46	2.21	21	1.78
22	6.06	4.61	3.73	3.14	2.88	2.52	2.26	22	1.82
23	6.06	4.61	3.74	3.15	2.96	2.59	2.32	23	1.86
24	6.06	4.62	3.74	3.15	3.06	2.67	2.38	24	1.90
25	6.06	4.62	3.74	3.16	3.16	2.74	2.44	25	1.94
26	0.06	4.62	3.74	3.16	3.26	2.82	2.50	26	1.98
27	6.07	4.62	3.75	3.17	3.37	2.91	2.57	27	2.03
28	6.07	4.62	3.75	3.17	3.50	3.00	2.64	28	2.08
29	6.07	4.63	3.76	3.18	3.62	3.09	2.72	29	2.13
30	6.07	4.63	3.76	3.19	3.76	3.19	2.80	30	2.18
31	6.07	4.63	3.77	3.20	3.91	3.30	2.88	31	2.24
32	6.08	4.64	3.78	3.21	4.08	3.42	2.97	32	2.30
33	6.08	4.64	3.79	3.23	4.26	3.54	3.06	33	2.36
34	6.09	4.65	3.80	3.25	4.45	3.67	3.16	34	2.42
35	6.09	4.66	3.81	3.27	4.66	3.81	3.27	35	2.49
36	6.10	4.68	3.83	3.30	4.90	3.97	3.38	36	2.56
37	6.11	4.69	3.85	3.33	5.16	4.14	3.51	37	2.63
38	6.12	4.70	3.88	3.37	5.45	4.32	3.64	38	2.70
39	6.13	4.72	3.91	3.41	5.77	4.52	3.78	39	2.78
40	6.14	4.74	3.94		6.14	4.74	3.94	40	2.86
41	6.16	4.77	3.98		6.59	4.98	4.10	41	2.94
42	6.18	4.80	4.02		7.10	5.25	4.28	42	3.02

Exercise

Calculate the monthly premium for each of the following policies:

	Type of policy	Profits	Age next birthday	Term	Sum assured
4	Whole life	'with'	24	—	£14 000
5	Endowment	'with'	30	25 years	£15 000
6	Endowment	'without'	36	20 years	£12 500
7	Endowment	'with'	42	to age 60	£12 400

Continue with Section K

K Progress check

Exercise

1 A lady invests £450 in a Finance Company offering an interest rate of $7\frac{1}{2}$%. How much interest should she receive at the end of a year?

2 A couple invest £800 in a Building Society. If the current interest rate is 9% per annum, how much will they receive after 6 months?

3 Find the compound interest on £1600 for 3 years at 5%.

4 A house is for sale at a price of £18 500. The Building Society's valuation is £18 000.

   (a) Calculate the total bill for the valuation using the scale of fees in Section F.
   (b) If the Building Society advance a 90% loan, calculate the amount of the loan and the deposit the buyer must provide.
   (c) If the loan is given for a period of 25 years at 11%, calculate the monthly repayments using the table in Section H.

5 Using the table in Section I, calculate the total annual insurance premium on a building valued at £26 000 and contents of £3000 covered comprehensively.

6 A lady aged 29 next birthday wishes to take out an endowment policy 'with profits' for £15 000, payments to be made over a 30 year period. Calculate her monthly premiums from the assurance tables provided in Section J.

Ask your teacher what to do next

L Inflation

The cost of most goods in the shops is increasing steadily.
This progressive increase is called **inflation**

Inflation ($+$) *adds* to the value

Example

As a result of general inflation, house prices increase by 7% in 1983, by 10% in 1984, and by 3% in 1985. How much would a house, bought for £17 000 at the beginning of 1983, be valued at the end of 1985?

Working

1983   Value at beginning = £17 000   1% of £17 000 = £170
Increase   = £ 1190 $\times 7$
Value at end   = £18 190   7% of £17 000 = £1190

1984   Value at beginning = £18 190   10% of £18 190 = £1819
Increase   = £ 1819
Value at end   = £20 009

1985   Value at beginning = £20 009   1% of £20 009 = £200.09
Increase   = £ 600 $\times 3$
Value at end   = £20 609   3% of £20 009 = £600.27
                                                 $\approx$ £600

The value of the house would be about £20 609 at the end of 1985.

Exercise

1   Due to inflation, the cost of a certain type of washing machine is increasing at a rate of 5% per annum. If its cost now is £320, what will the model cost two years from now?

2   The value of a house increased by 5% during 1979, by 8% during 1980 and by 10% during 1981. If a house cost £14 000 at the beginning of 1979, what would its value be at the end of 1981? (Round your final answer to the nearest £10.)

3   Due to inflation, the value of a painting is increasing at a rate of 8% per annum. If its value now is £1800, what will its value be in 2 years time? (Round your final answer to the nearest £100.)

Continue with Section M

M Depreciation

The value of most vehicles and machinery reduces with age.
This reduction is called **depreciation**.

Depreciation (−) *subtracts* from the value.

Example

The rate of depreciation of a £1000 knitting machine is reckoned to be 30% per annum. After how many years would its value fall below £350?

Working

1st year	Value	= £1000	10% of £1000	= £100
	(−) Depreciation	= £ 300		× 3
	Value at end of year	= £ 700	30% of £1000	= £300
2nd year	Value	= £ 700	10% of £700	= £ 70
	(−) Depreciation	= £ 210		× 3
	Value at end of year	= £ 490	30% of £700	= £210
3rd year	Value	= £ 490	10% of £490	= £ 49
	(−) Depreciation	= £ 147		× 3
	Value at end of year	= £ 343	30% of £490	= £147.00

Its value would fall below £350 after 3 years.

Exercise

1 A man bought a boat for £2000. If its value depreciates at 10% per annum, calculated on the value at the end of the previous year, what will its value be after 3 years.

2 A television set depreciates at the rate of 20% per year. If a new set costs £300, what will the set be valued at after 2 years?

3 A new car depreciates by 30% during its first year and then by 20% each year thereafter. If a car was bought from the factory for £8000, what would you expect to be the value of the car in three years time?

4 On making its annual report a company has to show the value of its machinery. If in 1982 this was valued at £40 000 and the depreciation is $6\frac{1}{2}$% per year, what would be the valuation in 1984?

Continue with Section N

N Loan repayment by instalments

The most usual method of paying back a loan is by instalments of a fixed amount (for example, £10 per month, £32 per 6 months, £50 per year). The borrower is usually charged interest on the amount outstanding during the period before each repayment.

Example

A man borrowed £500 from a Finance Company with an agreement to repay £100 each year. Interest of 8% was to be charged on the amount outstanding each year.
Calculate the amount of the loan which is outstanding after 3 repayments.

1st year

Loan = £500   1% of £500 =   £ 5
                                × 8
                8% of £500 =   £40

Interest = £40
Loan + Interest = £540

Repayment = £100

Amount still due = £440

2nd year	Loan	= £440.00	1% of £440 = £ 4.40
	Interest	= £ 35.20	× 8
	Loan and Interest	= £475.20	8% of £440 = £35.20
	Repayment	= £100.00	
	Amount still due	= £375.20	

Calculate interest on whole £s only.

3rd year	Loan	= £375.20	1% of £375 = £ 3.75
	Interest	= £ 30.00	× 8
	Loan and Interest	= £405.20	8% of £375 = £30.00
	Repayment	= £100.00	
	Amount still due	= £305.20	

Amount of loan outstanding after 3 repayments is £305.20.

Notice that the Amount still due is getting smaller each year.

Exercise

1 A man borrows £200 from a Loan Company and agrees to pay back the loan at £50 per annum. Interest was to be charged at 8% on the amount outstanding after each yearly payment. Calculate the amount outstanding after three repayments.

Copy and complete:

Working

1st year	Loan	= £200.00	1% of £200 = £ 2.00
	Interest	= £ 16.00 ◀	× 8
	Loan and Interest	= £216.00	8% of £200 = £16.00
	Repayment	= £ 50.00	
	Amount still due	= £166.00	

2nd year	Loan	= £166.00	1% of £166 = £ 1.66
	Interest	= £ 13.28 ◀	× 8
	Loan and Interest	= £179.28	8% of £166 = £13.28
	Repayment	= £▨▨▨	
	Amount still due	= £▨▨▨	

> Calculate interest on whole £s only.

3rd year	Loan	= £▨▨▨	1% of £▨▨ = £▨▨▨
	Interest	= £▨▨▨	× ▨
	Loan and Interest	= £▨▨▨	8% of £▨▨ = £▨▨▨
	Repayment	= £▨▨▨	
	Amount still due	= £▨▨▨	

2 A man borrows £300 from a Finance Company and agrees to repay it at £80 per annum. Calculate the amount outstanding after two repayments, interest being charged at $7\frac{1}{2}$% per annum.

Copy and complete:

Working

1st year	Loan	= £300.00	1% of £300 = £ 3.00
	Interest	= £ 22.50	× 7
	Loan and Interest	= £322.50	7% of £300 = £21.00
	Repayment	= £ 80.00	$\frac{1}{2}$% of £300 = £ 1.50
	Amount still due	= £242.50	£22.50

> Calculate interest on whole £s only.

2nd year	Loan	= £▨▨▨	1% of £▨▨ = £▨▨▨
	Interest	= £▨▨▨	
	Loan and Interest	= £▨▨▨	
	Repayment	= £▨▨▨	
	Amount still due	= £▨▨▨	

Continue with Section O

⊙ Quarterly and other repayments

Exercise

1 To buy a new stereo a man borrowed £180 agreeing to repay it at £40 per quarter (year). If the interest charged was 6% per annum on the amount outstanding in each quarter, calculate the amount outstanding after three repayments.

Copy and complete:

Working

1st quarter	Loan	= £180.00		1% of £180	= £ 1.80
	Interest	= £ 2.70 ◄			×6
	Loan and Interest	= £182.70		6% of £180	= £10.80
	Repayment	= £ 40.00		For a quarter	= £ 2.70
	Amount still due	= £142.70			

2nd quarter	Loan	= £142.70		1% of £142	= £ 1.42
	Interest	= £▢ ◄			×6
	Loan and Interest	= £▢		6% of £142	= £▢
	Repayment	= £▢		For a quarter	= £▢
	Amount still due	= £▢			

3rd quarter	Loan	= £▢			
	Interest	= £▢			
	Loan and Interest	= £▢			
	Repayment	= £▢			
	Amount still due	= £▢			

2 A woman borrowed £480 to buy a violin with an agreement to repay the loan at £120 half-yearly. If the interest is charged at $8\frac{1}{2}$% per annum on the amount outstanding in each half-year, calculate the amount still due after two repayments.

Copy and complete:

Working

1st half-year	Loan	= £480.00		1% of £480	= £ 4.80
	Interest	= £ 20.40 ◄			×8
	Loan and Interest	= £500.40		8% of £480	= £38.40
	Repayment	= £120.00		$\frac{1}{2}$% of £480	= £ 2.40
	Amount still due	= £380.40			= £40.80
				For half-year	= £20.40

2nd half-year	Loan	= £380.40		1% of £380	= £▢
	Interest	= £▢			
	Loan and Interest	= £▢			
	Repayment	= £▢			
	Amount still due	= £▢			

3 (a) A man borrowed £400 from a Finance Company with an agreement to repay £120 each year. Interest of 10% was to be charged on the amount outstanding. Calculate the amount of the loan which is outstanding after 2 yearly payments.

(b) If the loan was repaid at £60 each half-year, calculate the amount of the loan outstanding after two years (that is, after four half-yearly payments).

(c) If the loan was repaid at £10 each month, would the amount of the loan outstanding after 2 years be more or less than in parts (a) and (b)?

Continue with Section P

P Stocks and shares

The Stock Exchange is a market where stocks and shares are bought and sold.

Stock is a loan given to the Government, or a company, or a nationalized industry, or a regional council. If you buy stocks of this kind you receive interest at a fixed rate.

For example – $3\frac{1}{2}\%$ War Loan, 5% British Treasury Stock.

Example

What annual income would a person holding £550 of $3\frac{1}{2}\%$ War Loan receive?

$$\text{Income} = 3\frac{1}{2}\% \text{ of } £550 = £19.25$$

Working

$$1\% \text{ of } £550 = £\ 5.50$$
$$\times 3$$
$$3\% \text{ of } £550 = £16.50$$
$$\frac{1}{2}\% \text{ of } £550 = £\ 2.75$$
$$3\frac{1}{2}\% \text{ of } £550 = £19.25$$

Exercise

1   What annual income would a person holding £750 of 5% British Treasury Stock receive?

2   If I hold £1200 in $3\frac{1}{2}\%$ War Loan, what income should I receive each year?

Shares are issued by companies to raise money to expand business. Members of the public who buy shares in a company receive their share of the profits called a **dividend**.

New shares are bought at a fixed price such as 25p, 50p or £1.

For example, ICI £1 shares, Guinness 25p shares, Dunlop 50p shares.

Example

Cadbury Schweppes, a company which issues 25p shares, declared a dividend of 3p per share. If I hold 1800 shares, what dividend should I expect to receive?

$$\text{Dividend per share} = 3p$$
$$\text{Dividend from 1800 shares} = 1800 \times 3p$$
$$= 5400p$$
$$= £54$$

Exercise

3   The Scottish and Newcastle, a brewery company issuing 20p shares, declared a dividend of 2.4p per share. If a man holds 800 shares in the company, what dividend should he expect to receive?

4   Distillers have declared a dividend of £0.075 per share. If I hold 4000 shares, what dividend should I receive?

5   Tate and Lyle £1 shares pay a dividend of £0.135 per share. How much should a lady holding 850 shares receive?

Continue with Section Q

Q Yield on shares

Stocks and shares may be sold through the Stock Market by one person to another. If a company is doing well and paying a high rate of interest, the buyer might be willing to pay a high price for each share.

However, if a company is having difficulties and has been paying a low rate of interest, the buyer will probably be able to purchase shares at a low price.

To make a wise investment, the investor should first of all calculate the **yield**, that is, the expected percentage return on the money he invests.

Example

Tarmac shares are quoted at 88p and pay a dividend of 10p per share.
(a) What does it cost to buy 300 shares?
(b) What dividend should be expected?
(c) What is the yield as a percentage of the money invested?

(a) Shares are quoted at 88p, that is, cost of 1 share = 88p
Cost of 300 shares = $300 \times 88p = £264$

(b) Dividend = $300 \times 10p = £30$

(c) The yield means the total amount of dividend received on the shares, so yield = £30.
The yield as a fraction of the price = $\frac{30}{264} = 11.4\%$.

Exercise

(Give percentage yields to one decimal place.)

1 Glaxo shares are quoted at £1.80 and pay a dividend of 11p per share.
(a) What does it cost to buy 400 shares?
(b) What dividend should be expected?
(c) What is the yield as a percentage?

2 Bank of Scotland shares have a market value of £1.28 and the dividend is 10p per share.
(a) What does it cost to buy 600 shares?
(b) What dividend should be expected?
(c) What is the yield as a percentage?

3 How many Tesco shares quoted on the Stock Exchange at 28p can I buy for £100? Give your answer as a whole number of shares.

4 (a) How many EMI shares quoted at 70p can I buy for £210?
(b) If a dividend of 4p per share is declared, what income should be expected?
(c) What is the percentage yield?

5 I have £48 to invest. Which investment provides the greater yield
(a) Granada shares quoted at 30p declaring a dividend of 3p per share
(b) STV shares quoted at 16p declaring a dividend of 2p per share?

6 A man spends £1120 in buying shares at 70p each and sells them again when they rise to 88p. How much does he gain?

Continue with Section R

R Comparison of investments

A man had £1000 to invest.

The graph shows how the value of his money might have changed, depending on how he invested it, and also shows the increase in the cost of living over 4 years.

How your money grows!

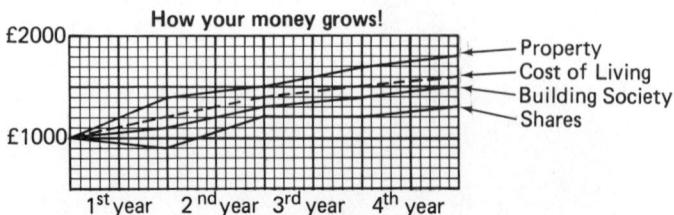

Exercise

1   From the graph, find the value of each of the three investments at the end of the fourth year.

2   The dotted line shows the rise in the COST OF LIVING. From the graph, state how much goods purchased for £1000 would cost at the end of the fourth year.

3   What investments (a) had kept their value above the rise in the cost of living and (b) had fallen behind?

4   What is the percentage increase in the cost of living over the four years?

Continue with Section S

S Progress check

1   A house was bought in March 1983 for £17 500. By March 1984 house prices had increased by 8%, by March 1985 they had increased a further 5%, and by March 1986 a further 10%. Find the value of the house in March 1986. (Round the final answer to the nearest £10.)

2   A loan of £200 is to be repaid by instalments of £50 per half year. The interest is 12% per annum and is charged on the amount outstanding in each half year. Calculate the amount still due after two repayments.

3   National Commercial shares are quoted at 68p and pay a dividend of 5.1p per share
    (a) What does it cost to buy 400 shares?
    (b) What is the expected dividend?
    (c) What is the yield as a percentage?

Tell your teacher you have finished this unit

UNIT 9 Pictorial Representation

A Pictograms

Population Densities of E.E.C. Countries

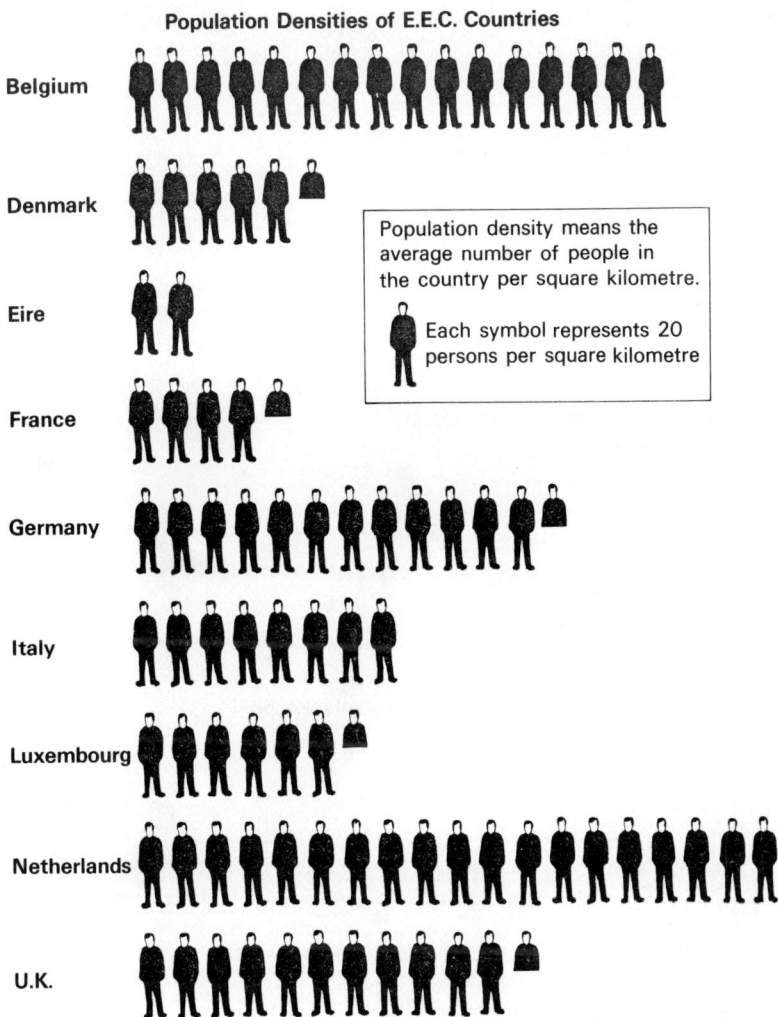

Belgium

Denmark

Population density means the average number of people in the country per square kilometre.

Each symbol represents 20 persons per square kilometre

Eire

France

Germany

Italy

Luxembourg

Netherlands

U.K.

The illustration above is an example of a **pictogram**.

This is a way of providing a quick impression of statistical information.

Its disadvantage is that it is not possible to read from the pictogram precisely.

Notice that a pictogram must have a heading to tell what it presents and an explanation of what each symbol represents.

Exercise

1 Which of the E.E.C. countries is most densely populated?

2 Which of the E.E.C. countries is least densely populated?

3 Is the population density greater or less in Germany than in France?

Example

To find the population density for Denmark
we count the number of symbols.
You can see that there are five and a half (5.5).

Each symbol represents 20 people.

So the population density is

$5.5 \times 20 = 55 \times 2$

$= 110$ persons per square kilometre.

Denmark

Exercise

4 What is the population density of (a) Belgium and (b) France?

5 Which of the following is the population density in persons per square kilometre for Luxembourg?

A 5.5   B 6.5   C 110   D 130   E 650

6 Can you tell *from the pictogram* whether there are more or fewer people in Germany than in France? Give a reason for your answer.

7 The population densities (in persons per square kilometre) of 7 European non–E.E.C. countries are listed below:

Spain Portugal Hungary Poland Switzerland Austria Greece
70   100   110   100   150   90   70

Draw a pictogram like the one on page 165 to show this information.

Sometimes you see a pictogram in which more than one symbol is used.

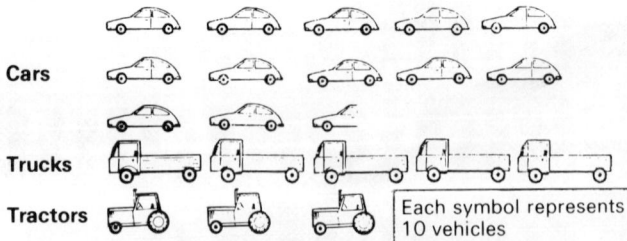

WEIR MOTORS – Vehicles Sold

Cars

Trucks

Tractors

Each symbol represents 10 vehicles

8 For which type of vehicle was the greatest number sold?

9 For which type of vehicle was the smallest number sold?

10 How many tractors were sold?

11 Approximately how many cars were sold?

12 A traffic census was taken between 9.30 a.m. and 10 a.m. and gave the following results:

Bicycles 10   Cars 25   Lorries 15   Buses 8

Draw a pictogram showing these results using one symbol to represent 5 vehicles.

Continue with Section B

B Bar charts

Houses Built in Weir New Town 1982–1985

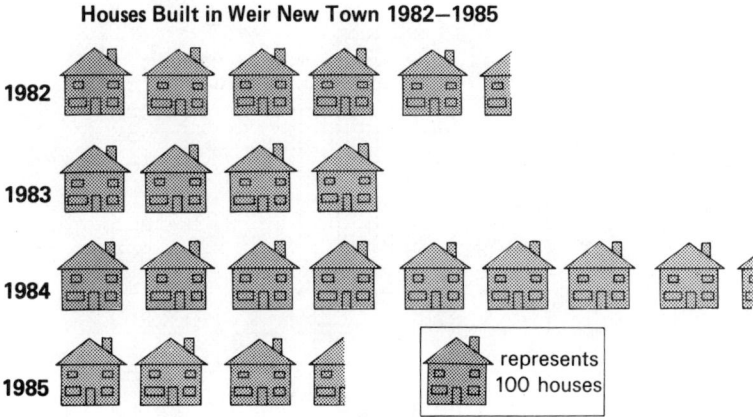

The pictogram above shows that in 1982 there were more than 500 and less than 600 houses built in Weir New Town. We might guess the number to be about 550.

If we take the pictogram and, for each year, move all the houses close together and remove their roofs we get this:

Houses Built in Weir New Town 1982–1985

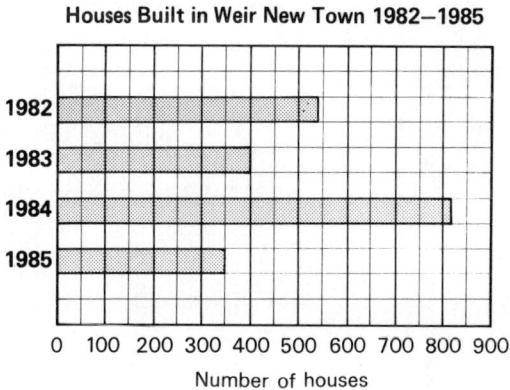

Number of houses

The result is called a **bar chart**.

Although the information is not in such a striking form as in a pictogram, it is possible to get more accurate information from it. Just as accurately as you can measure in fact, which in this case should be to the nearest 10 houses.

We can now tell that the number of houses built in 1982 was about 540.

Bar charts can be drawn with the bars horizontal as above, or vertical as in the next example.

Exercise

1 From the bar chart on page 167 find as accurately as you can the number of houses built in each of the years (a) 1983, (b) 1984, (c) 1985.

2 In 1986, 650 houses were built. Draw a bar chart showing the houses built in Weir New Town 1983–1986.

Presentations of Candidates at 1984 Examinations

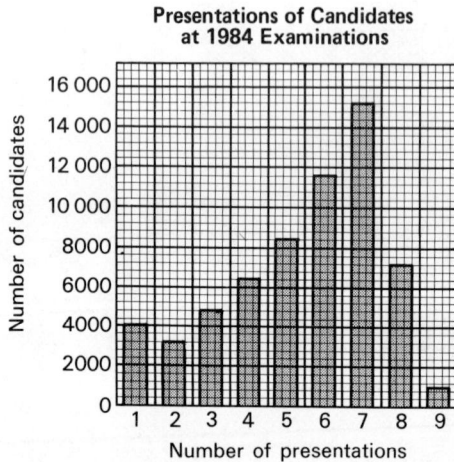

If in an examination you sit two subjects – say English and Mathematics – then you are said to have 'two presentations'.

The diagram above shows the number of candidates who were presented in 1, 2, 3, . . ., 9 subjects.

For example, you can see that about 4000 candidates were presented in 1 subject and about 3200 in 2 subjects.

Notice that each bar must be labelled to show what it refers to – in the first example the year 1982, 1983, 1984, or 1985 and in the second the number of presentations 1, 2, 3 . . ., 8 or 9. Notice also that there must be a scale on the other axis.

Exercise

3 Is the number of candidates presented in 4 subjects greater or less than the number presented in 8 subjects?

4 What number of presentations were made by the greatest number of candidates?

5 Find the number of candidates presented in 5 subjects.

6 How many candidates were presented in fewer than 4 subjects?

7 How many candidates were presented in 7 or more subjects?

Sometimes more than one bar chart is drawn in the same diagram.

Age Distribution of U.K. Population

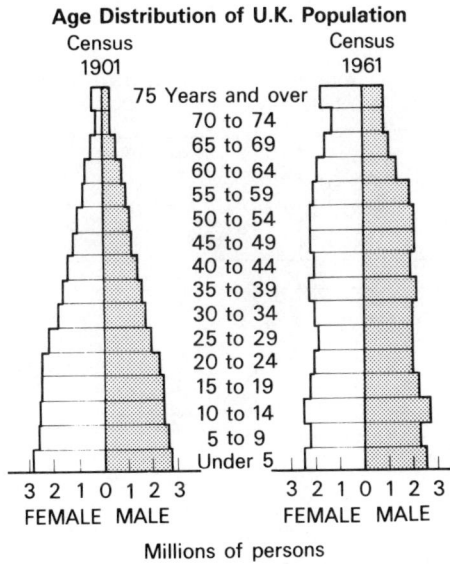

Census 1901   Census 1961

75 Years and over
70 to 74
65 to 69
60 to 64
55 to 59
50 to 54
45 to 49
40 to 44
35 to 39
30 to 34
25 to 29
20 to 24
15 to 19
10 to 14
5 to 9
Under 5

3 2 1 0 1 2 3
FEMALE MALE

3 2 1 0 1 2 3
FEMALE MALE

Millions of persons

Exercise

8 In 1961 were there more females than males aged 75 years and over?

9 In 1901 were there more females than males aged 70 to 74?

10 Were there more children under 5 years old in 1901 than in 1961?

11 Were there more people aged 75 years and over in 1901 than in 1961?

12 Did the total U.K. population increase or decrease between 1901 and 1961?

13 The shape of each distribution is quite different. Which of the following explanations is most likely?

    A. The birth rate was lower in 1901.
    B. People are living longer in 1961.
    C. 1901 was an unusual year.

Continue with Section C

C Bar chart used for comparison

Value of Imports to Franconia 1976 and 1986
Distributed by types of transport

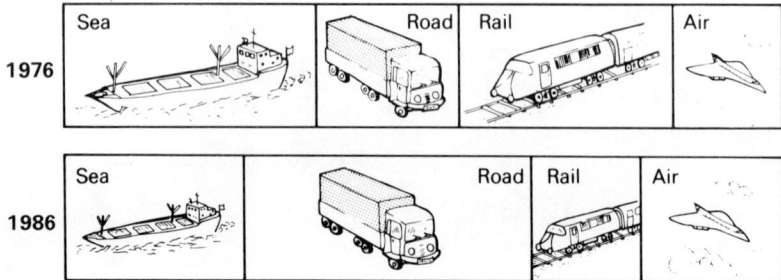

Example

Find the percentage of imports which came by sea in 1976.

Measure the length of the bar representing sea transport.
It is 3.5 cm.
Measure the length of the whole bar (which represents the total imports).
This length is 10 cm.

So the fraction coming by sea is $\dfrac{3.5}{10} = \dfrac{35}{100} = 35\%$.

Exercise

1 (a) In 1986 what percentage of imports came by sea?
(b) In 1976 what percentage of imports came by air?
(c) In 1986 what percentage of imports came by air?

Example

If in 1986 imports by road were valued at £220m, what was the value of imports by sea in 1986? (Note: £200m is another way of writing £200 000 000.)

Measure the lengths of the bars representing road and sea imports in 1986.

	cm	£

Road imports are represented by 4 cm.
Sea imports are represented by 2.5 cm.

$$4 \longrightarrow 200\,000\,000$$
$$1 \longrightarrow 50\,000\,000$$
$$2.5 \longrightarrow 2.5 \times 50\,000\,000$$
$$= 125\,000\,000$$

So, value of imports by sea was £125m.

Exercise

1 (d) Which types of transport increased their share of imports in 1986 compared with 1976?
(e) From the above chart alone can you tell whether the value of imports by sea to Franconia increased or decreased between 1976 and 1986? Give a reason for your answer.
(f) If in 1976 rail imports were valued at £90m what was the value of sea imports?

2

Fuel used in Houses in 1979 and 1983

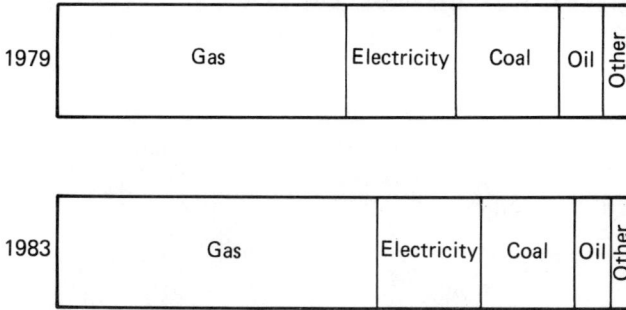

| 1979 | Gas | Electricity | Coal | Oil | Other |

| 1983 | Gas | Electricity | Coal | Oil | Other |

The graph above shows how the various fuels were used in 1979 and 1983. Each type of fuel is shown as a fraction of the total fuel used.

(a) Which type(s) of fuel increased its share from 1979 to 1983?
(b) By measuring the lengths of the rectangles find the percentage share of coal in 1979 and 1983.
(c) If the total consumption of fuel in homes in 1983 was 15 500 million therms, find the actual consumption of electricity.
(d) Can you explain why the gas industry's share increased between 1979 and 1983?

3 The bar chart shows the diameters of a sample of Scotlon fibres in millimetres.

(a) What diameter occurred most frequently in this sample?
(b) How many fibres had a diameter of 0.13 mm?
(c) How many fibres were tested altogether?
(d) What percentage of the fibres had a diameter of *less than* 0.15 mm?
(e) If a diameter of 0.19 mm or more is considered too coarse, what percentage of the fibres was unsatisfactory?

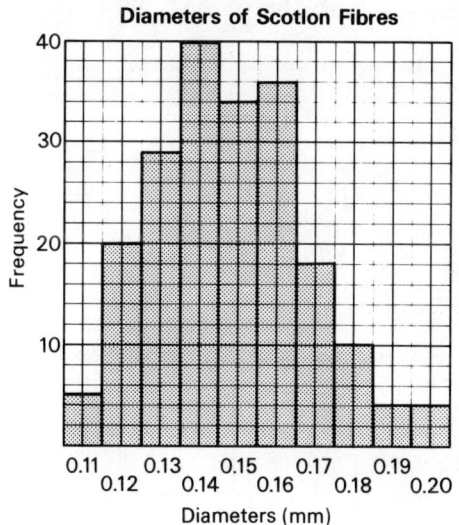

Diameters of Scotlon Fibres

Continue with Section D

D Bar charts with broken scale

The table below shows the number of primary school pupils in Scotland in the years 1980 to 1983.

Year	1980	1981	1982	1983
Pupils (thousands)	552	525	499	474

If a bar chart were drawn of this information it would look like this:

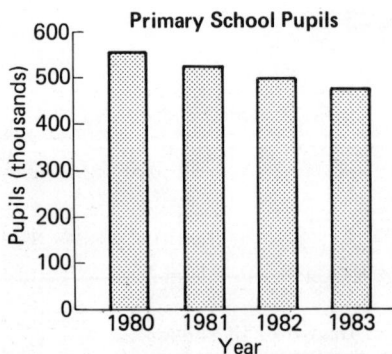

The decrease over the years does not stand out well.

In order to emphasise the differences, bar charts are sometimes drawn where the bottom part has been removed and the remainder expanded as below. This bar chart represents the same information as the one above.

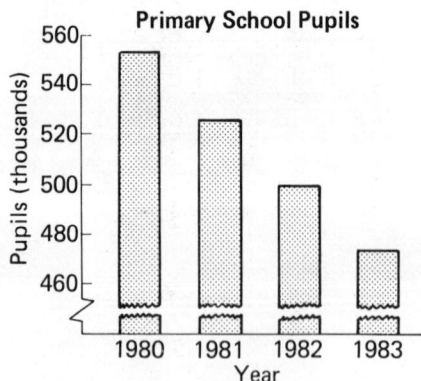

Notice that a break has been shown on the vertical scale and also on each bar. (The break in the bars is often omitted.)

Exercise

1 (a) Which of the months shown has most sunshine?
  (b) Which month has least sunshine?
  (c) What is the average daily hours of sunshine in
   (i) December
   (ii) March?
  (d) On average how many more hours of sunshine are there in April than in January?
  (e) Would you expect San Matilda to be North or South of the equator? Give a reason for your answer.

Average Daily Hours of Sunshine in San Matilda

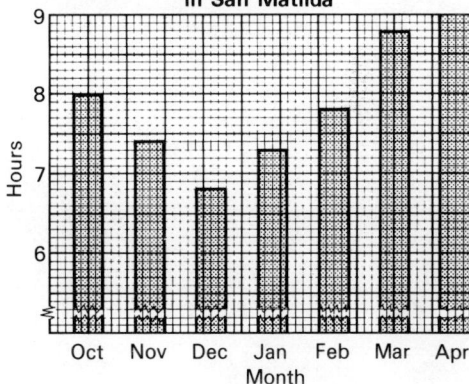

2 The diagram opposite shows the monthly sales of a product. The production costs were constant at £8500 per month.

  (a) During which month were sales highest?
  (b) Calculate the profit made in February.
  (c) Calculate the *Total* profit for the half year.
  (d) Express the profit made in February as a percentage of the total profit for the half year.

Monthly Sales

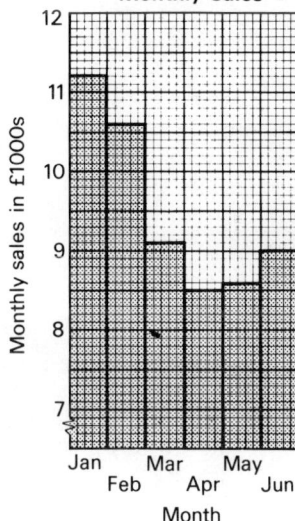

Continue with Section E

E Histograms

A **histogram** is similar to a bar chart. Here is a histogram.

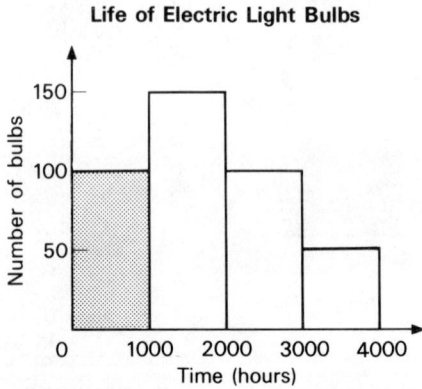

Life of Electric Light Bulbs

Notice that

1. The 'bars' are joined together, not separated as in the example on page 172.

2. The horizontal scale is shown as a continuous scale.

The first rectangle (shaded) represents the number of bulbs in the sample which lasted less than 1000 hours. Strictly it is the area of the rectangle which represents this number but since all the histograms you are likely to meet will be made of rect-angles with the same width, you can take the height as representing the total number. From the histogram 100 bulbs lasted less than 1000 hours.

Exercise

1 (a) How many bulbs in the sample have lives between 1000 and 2000 hours?
   (b) How many bulbs in the sample have lives between 2000 and 3000 hours?
   (c) How many bulbs in the sample have lives between 3000 and 4000 hours?
   (d) If we assume the 100 bulbs lasting less than 1000 hours have an average life of 500 hours, what would be the total of the life times of these bulbs?
   (e) Assuming average lives of 1500, 2500, and 3500 hours for each of the next three classes, what are the total life times for each of these groups?
   (f) What is the total of life time of all the bulbs?
   (g) What is the total number of bulbs shown in the histogram?
   (h) What is the average life time of a bulb in this sample?

2 The histogram opposite shows the lengths in millimetres of a sample of 200 leaves.

Sample of Leaves

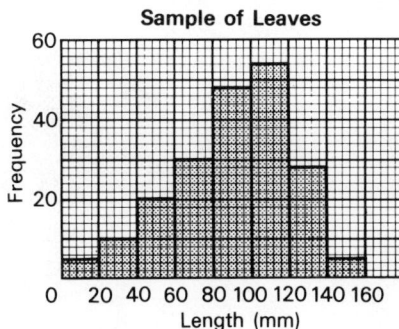

(a) Copy and complete the following table:

Length (mm)	0–20	20–40	40–60	60–80	80–100	100–120	120–140	140–160
Frequency	5							

(b) Check that the total of the numbers in the frequency row is 200.
(c) Assuming the average length of those leaves having lengths in the range 0–20 mm is 10 mm, what is the total of the lengths of the leaves in this class?
(d) Copy and complete the following table:

Length (mm)	Mid-point, X	Frequency, f	$f \times X$
0–20	10	5	50
20–40	30		
40–60			
60–80			
80–100		48	
100–120			
120–140			
140–160	150	5	750

This means lengths greater than or equal to 40 but *less* than 60. (points to 40–60)

(e) Find the sum of the numbers in the $f \times X$ column.
(f) Divide the answer to (e) by the answer in (b) to find the average length of a leaf. (Give answer correct to the nearest 10 mm.)

3 Draw a histogram to show the distribution of heights of 100 pupils in a year group shown in the following table:

Height (cm)	156–158	158–160	160–162	162–164	164–166	166–168
Frequency	2	8	36	40	12	2

Continue with Section F

F   Pie charts – interpretation

This diagram is called a **pie chart**. It is another way of illustrating statistical information.

The area of the whole circle represents the total number of houses built and the areas of the parts represent the numbers of the different types of house.

From the diagram you can tell at once that more than half the houses had three bedrooms and about a quarter had two bedrooms.

New Houses Built 1986

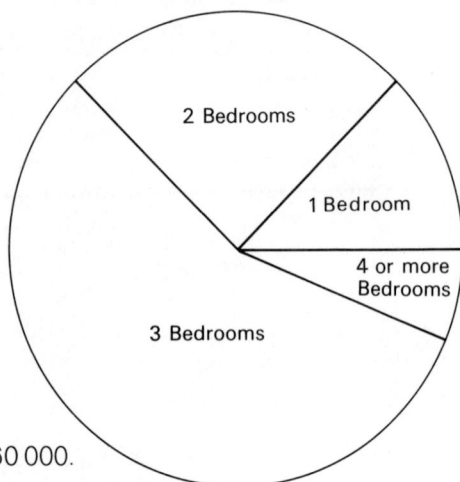

Total number of houses = 360 000.

If you want to find the numbers more accurately you would need to measure the angles at the centre of the circle.

Example

To find the fraction of houses with 2 bedrooms, measure the size of the shaded angle with a protractor.

You should find that it is 90°.

Remembering that a complete rotation is 360°, this means that the fraction of houses with 2 bedrooms is $\dfrac{90}{360} = \dfrac{1}{4} = 25\%$.

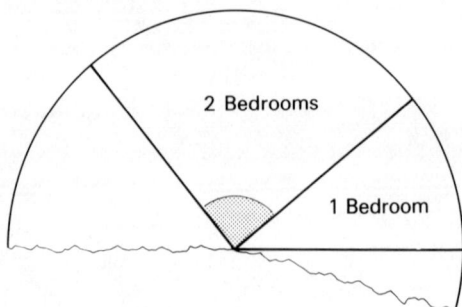

To find the number of houses with 2 bedrooms:

We know that 360° represents 360 000 houses

so   10° represents 360 000 ÷ 36 = 10 000 houses

so   90° represents 9 × 10 000   = 90 000 houses

Exercise

1 The pie chart shows the destinations of school leavers with 1 or more Higher grade passes.
   (a) If the total number of such leavers was 18 000, how many went to Further Education?
   (b) How many leavers entered Employment?
   (c) How many leavers entered Nursing?

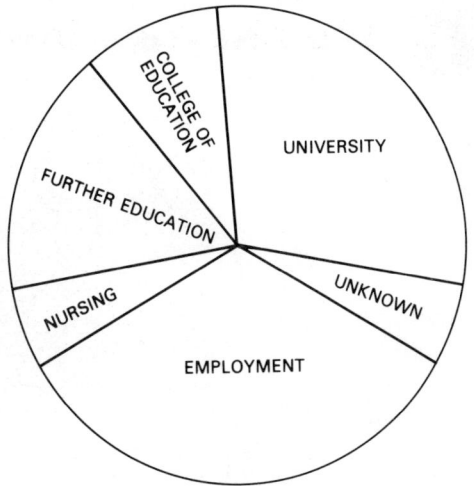

2 The pie chart shows the way in which a company's total manufacturing costs of £240 000 for a year were made up.
   (a) What percentage of the total costs was represented by packaging and sales expenses together?
   (b) What was the cost of wages and salaries?
   (c) Which two costs were equal?

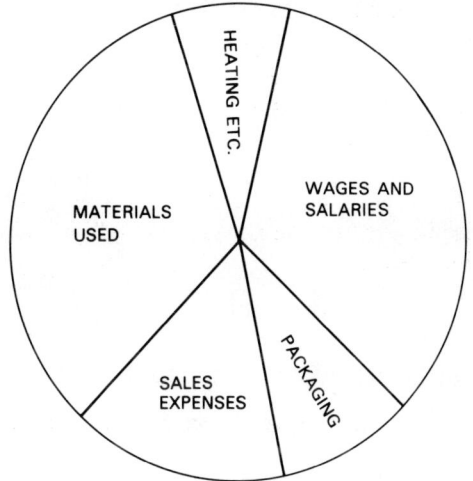

The pie chart shows the country of origin of the 7 200 000 visitors to the U.K.
   (a) Which country did the greatest number of visitors come from?
   (b) Which West European country gave rise to the greatest number of visitors?
   (c) Approximately how many visitors to the U.K. came from Canada?
   (d) Name one country from which the visitors would be included in the 'Others' sector?
   (e) If a visitor was picked at random, what is the probability that that person was from one of the countries included in 'Other West European' and 'Others'?

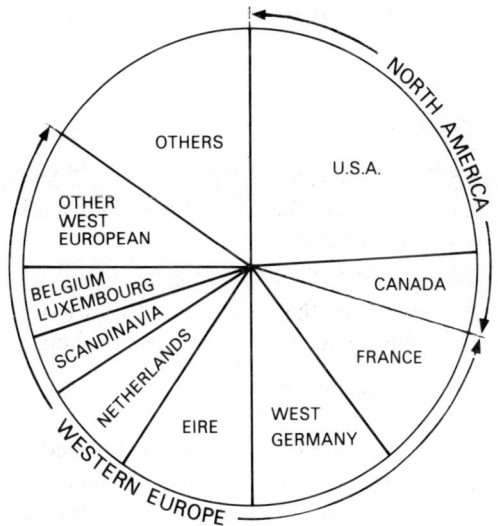

Continue with Section G

G Pie charts – construction

A sixth year pupil's 40–period week is divided up as follows:

Maths	12 periods
Physics	8 periods
Chemistry	8 periods
Study	6 periods
Other	6 periods
Total	40 periods

To show this information on a pie chart we work out what angle represents each subject.

40 periods is represented by 360°

so   1 period is represented by $360° \div 40 = 9°$

so   12 periods is represented by $12 \times 9° = 108°$

so the Maths periods are represented by a sector of angle 108°.

We draw a suitable size of circle and at the centre make an angle of 108° thus:

We then label the sector Maths.

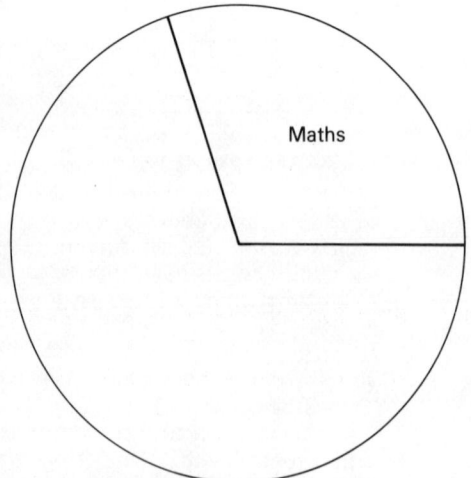

The angle for Physics will be $8 \times 9° = 72°$

Starting from the edge of the Maths sector we make an angle of 72° and label the sector Physics.

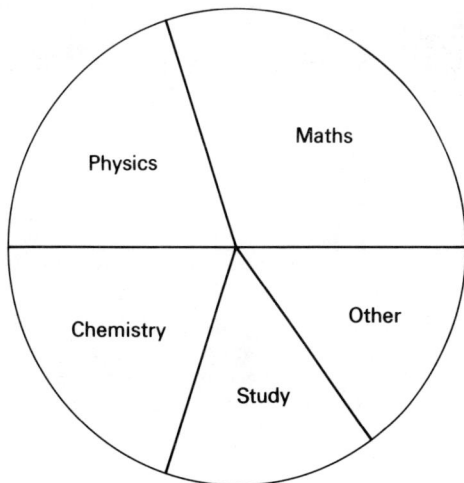

Continue in this way until the whole circle has been divided up.

Exercise

1 A pupil draws a pie chart to show how he spends his day. He sleeps for 8 hours. What angle should be at the centre of the sector showing sleep?

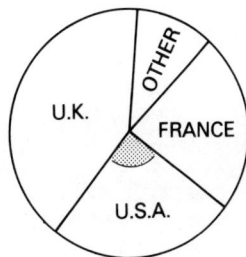

2 The pie chart shows the country of origin of 100 guests in a hotel. If 25 came from the U.S.A., what is the size of the shaded angle?

3 A girl spends $\frac{1}{2}$ her pocket money on records, $\frac{1}{3}$ on sweets, and $\frac{1}{6}$ on other purchases. Show this on a pie chart.

4 A survey of types of cars using a car park is made with the following results:

BL	Ford	Vauxhall	Other	Total
120	80	60	100	360

Show this information on a pie chart.

5 In 1986, 30 000 pupils were enrolled in secondary schools in an African State. The distribution of pupils among the four provinces was as follows:

Northern 2500,   Southern 6000,   Western 20 000,   Eastern 1500.

Show this information on a pie chart.

Continue with Section H

H Progress check

Exercise

1 A girl is making a pie chart to show her school subjects. If one–fifth of her time is spent on shorthand and typing, what angle should be at the centre of the sector showing this?

2 The bar chart shows the number of passes obtained by a random sample of 25 exami-nation candidates from a certain school.

(a) How many candidates obtained exactly three passes?

(b) What was the total number of passes obtained by those who obtained exactly three passes?

(c) What is the total number of passes gained by all the candidates in the sample?

Examination Passes

(d) How many candidates obtained three or more passes?

(e) What percentage of the candidates gained less than three passes?

(f) What is the probability that a candidate, selected at random from this sample, gained exactly four passes?

(g) If the school presented 165 candidates altogether, estimate the number who gained exactly four passes.

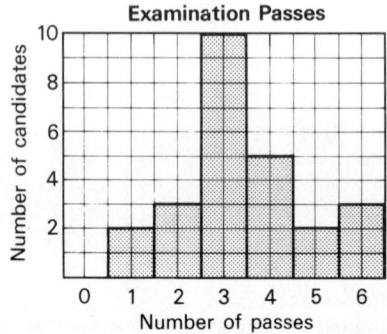

3 The bar graph shows an estimated in-crease in personal spending by 1990 compared with that in 1984.

(a) Give the total estimated expenditure on food in 1990.

(b) For which item is it estimated that there will be the greatest actual in-crease by 1990?

(c) Express the estimated increase in spending on housing in 1990 as a percentage of the spending in 1984.

(d) The graph shows that there is one pair of items, (i) and (ii), for which the expenditure on (i) was less than that on (ii) in 1984, but the estimate of expenditure on (i) in 1990 is more than on (ii). Which pair?

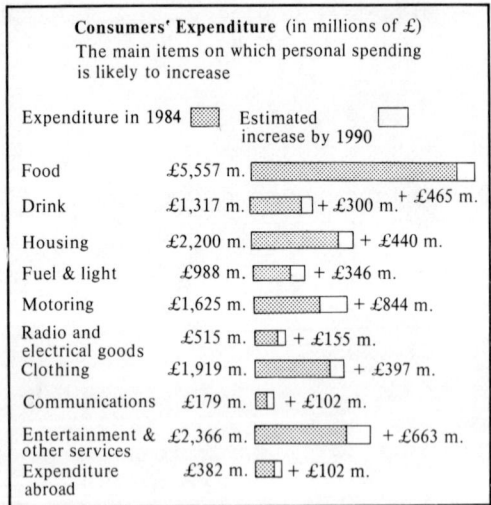

Consumers' Expenditure (in millions of £)
The main items on which personal spending is likely to increase

Expenditure in 1984 ▨   Estimated increase by 1990 ☐

Food	£5,557 m.	+ £465 m.
Drink	£1,317 m.	+ £300 m.
Housing	£2,200 m.	+ £440 m.
Fuel & light	£988 m.	+ £346 m.
Motoring	£1,625 m.	+ £844 m.
Radio and electrical goods	£515 m.	+ £155 m.
Clothing	£1,919 m.	+ £397 m.
Communications	£179 m.	+ £102 m.
Entertainment & other services	£2,366 m.	+ £663 m.
Expenditure abroad	£382 m.	+ £102 m.

Ask your teacher what to do next

∎ Misleading pictograms

Sometimes you may see pictograms which are designed to mislead.

Misleading Example
Daily production of milk doubled between
1982 and 1985 at Weir Dairy Farm.

Weir Dairy Farm

Since the daily production of milk doubled between 1982 and 1985, the illustration has been drawn with all dimensions doubled.

So the scale factor is $k = 2$.

Looking at the picture the impression given is either the change in area, or the change in volume of bottles of this shape.

Scale factor for area is $k^2 = 2^2 = 2 \times 2 = 4$

Scale factor for volume is $k^3 = 2^3 = 2 \times 2 \times 2 = 8$

So the picture gives the impression that the production in 1985 is either 4 times or 8 times that in 1982.

Exercise

1 Draw a pictogram to correctly show doubled production at Weir Dairy Farm between 1982 and 1985.

Continue with Section J

J Interpretation of graphs

Another way of showing statistical information pictorially is shown below.
The graph has been drawn to show the percentage of unemployed and the percentage of vacancies between 1980 and 1983.
In interpreting such a graph you must look carefully at the two scales.

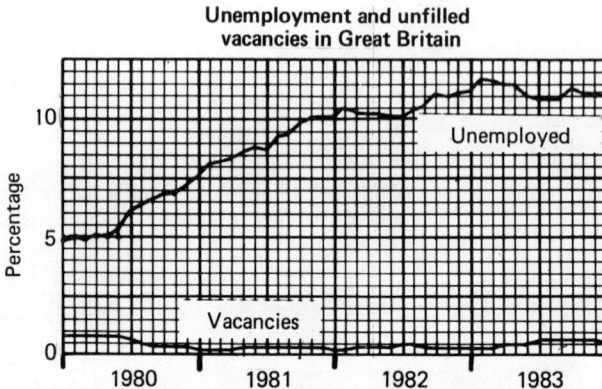

Unemployment and unfilled
vacancies in Great Britain

The vertical scale is marked off in percentages from 0 to 10. Each unit has been divided into 2 so each small square has a side representing 0.5.

The horizontal scale is marked off in years, each year divided into 12 so that the side of a small square represents one month. The dashes on the horizontal scale show January of each year.

To find the percentage unemployed in February 1981 we count along from the dash at the beginning of 1981 which represents the 1st of January and then read up the vertical scale to find the percentage as 8.2%.

Exercise

1   Use the graph above to answer the following:
   (a) In what month did unemployment first reach 10%?
   (b) In the period shown unemployment increased throughout 1980 and 1981. What was the first month in which it started to decline?
   (c) Between which two months in 1983 did unemployment fall?
   (d) In which year were the vacancies highest?

2 The graph shows the depth of the water in a tidal river from 6 am to 6 pm on a certain day. Thus, at time t, d represents the depth of water.

From the graph find, to the nearest 0.2 metre

(a) the value of d at 9 am,
(b) the distance between water level at 4.30 pm and a bridge 10 m above the bed of the river,

and, to the nearest 10 minutes,

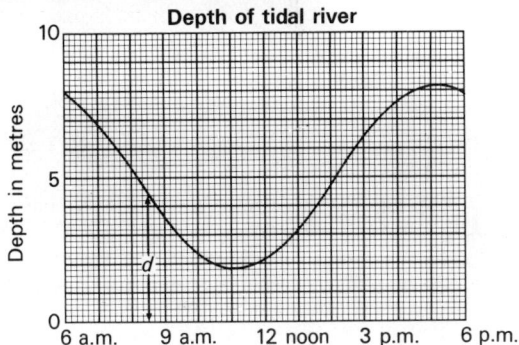

Depth of tidal river

(c) the earliest time after 6 am when a yacht, whose mast requires a clearance of 5.2 metres, can sail under the bridge,
(d) the times between which a barge requiring 2.4 metres depth of water cannot pass under the bridge.

3 Three substances are tested independently for solubility in water at different temperatures. The graph shows the results.

(a) What is the maximum weight of substance C which will dissolve in 100 g of water at 50°C?
(b) Which of the three substances is the most soluble at 20°C?
(c) At what temperature is the solubility of substance A the same as that of substance B?
(d) How much of substance A will dissolve in 100 g of water at 10°C?
(e) How much *more* of substance A will dissolve in 100 g of water at 70°C than at 10°C?
(f) At what temperature will the same amount of substances B and C dissolve in 100 g of water?

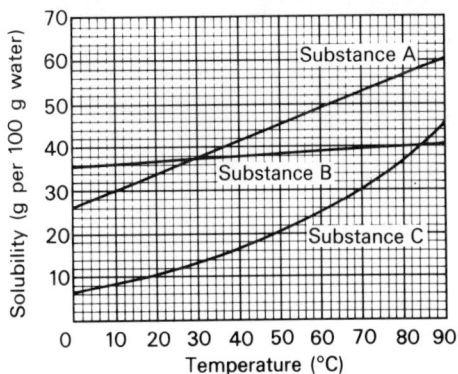

Solubility in Water at Different Temperatures

4 The graph shows how the price of a £5 basket of groceries has increased from 1979 to 1983. The prices are expressed to the nearest 10p. Notice the break in the vertical scale.

(a) Between which two years was the price increase greatest?
(b) A hotel bought £200 worth of groceries in 1979. How much would the same groceries cost in 1980?
(c) What was the total increase in the price of the £5 basket of groceries over the 4 year period. What was the average increase?
(d) Calculate the percentage increase between 1979 and 1980.

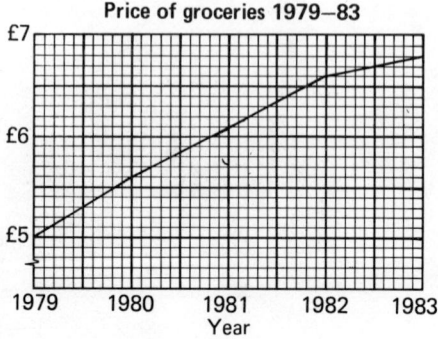

Price of groceries 1979–83

5 Each month the total number of kilometres travelled by a car since new was recorded on a graph. The part relating to the year 1985 is shown.

(a) How many kilometres had the car travelled by 31st Dec. 1985?
(b) How many kilometres had the car travelled by 31st Dec. 1984?
(c) Find the total number of kilometres travelled during the year.
(d) During which month did the car travel the greatest number of kilometres?
(e) Find the average number of kilometres travelled per month, to the nearest ten kilometres.
(f) If the year above is typical of the extent to which the car was used, calculate to the nearest month how old the car was at the beginning of 1985.

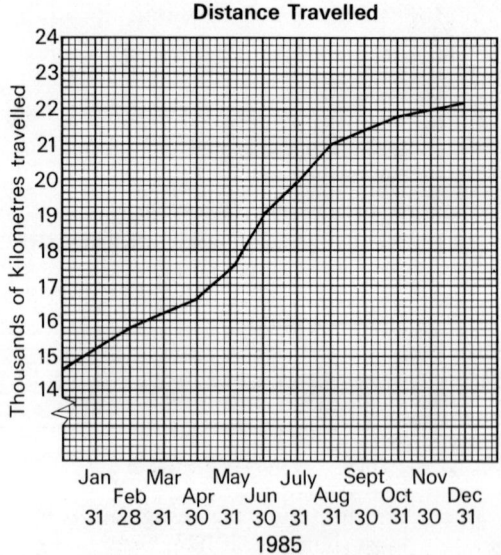

Distance Travelled

6

The graph shows the rate of a man's heartbeat and the total volume of blood that his left ventricle pumped out during each minute of a fifteen minute period.

During the five minutes shown by the thick black line along the time scale the man was exercising strenuously.

Notice that two vertical scales are shown. One on the left for heart rate and one on the right for output of blood.

Heart Rate and Volume of Blood

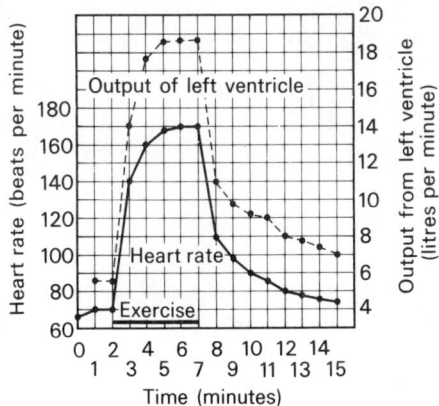

Using the information given by the graph, copy and complete the table below.

$$\text{Volume per beat} = \frac{5.6 \text{ litres}}{70 \text{ beats}}$$
$$= \frac{5600 \text{ ml}}{70 \text{ beats}}$$
$$= 80 \text{ ml per beat}$$

Time	Heart rate in beats per min	Output of left ventricle in litres per min	Volume of blood pumped out per beat in ml
immediately before exercise	70	5.6	80
after 1 min. of exercise			
... min. after end of exercise			
... min. after end of exercise			

7   **Trade Unions Membership**

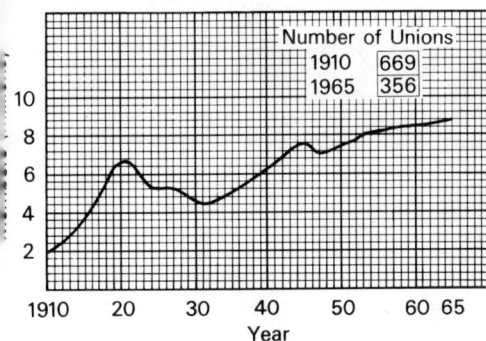

Number of Unions	
1910	669
1965	356

The diagram shows the number of Trades Union members between 1910 and 1965.

(a) How many union members were there (i) 1910 and (ii) 1965?

(b) Calculate the average number of members of a union in 1910 and in 1965. (Give answers to the nearest thousand members.)

(c) After 1921 the number of union members fell. In what year thereafter did it reach its lowest level?

(d) In which year did the total number of members first reach 8 million?

8 The graph shows the amount of petrol (in litres) in the tank of a car during a journey of 500 km.

(a) The tank was replenished twice during the journey. After what distances?

(b) How much petrol was used during the first 50 km of the journey?

(c) How much petrol was used over the whole journey?

(d) Rate of petrol consumption can be measured in litres per 100 km. What is the car's rate of petrol consumption on this journey?

(e) Is the rate of using petrol greater or less over the first 50 km than over the second 50 km?

Petrol Consumption

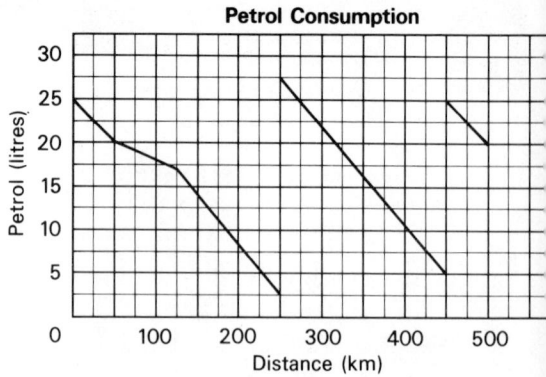

9 The full line graph shows the number of passengers carried by buses belonging to a city's transport department during a certain day, and the broken line graph shows the total number of seats available throughout the period.

(a) At what time were most passengers carried?

(b) Approximately how many passengers were being carried at any given time between 1430 and 1530 hours?

(c) At 1300 hours what percentage of the seating capacity was being used?

Comparison of Bus Passenger Use and Capacity

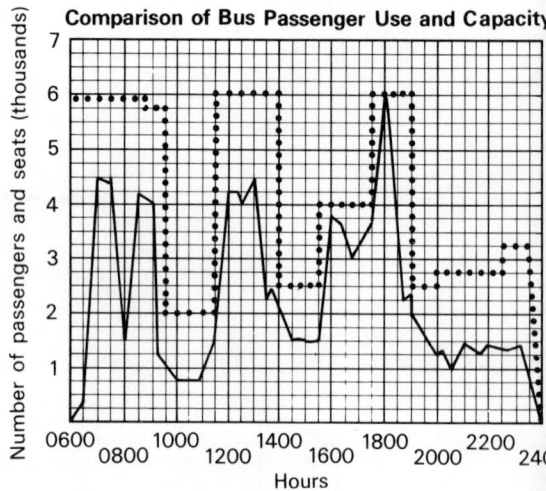

Continue with Section K

K Progress check

Exercise

1 The diagram shows the expected trends in travelling habits in Scotland, expressed in millions of passenger kilometres for each year. From the diagram answer the following questions:

(a) Which form of transport is expected to decrease in popularity?

(b) In which year did the car overtake the bus as a more popular form of transport?

(c) How many passenger kilometres are expected to be covered by car in 1984?

(d) In which year is the number of passenger kilometres travelled by car expected to reach 4000 million?

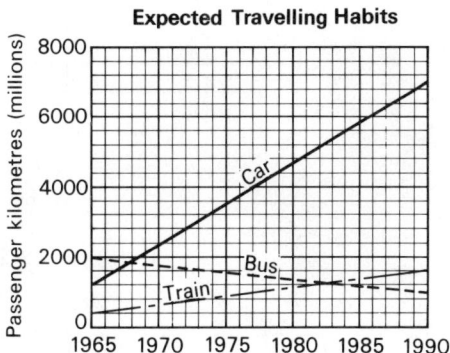

Expected Travelling Habits

2 A hiker leaves a hostel at 11 am to walk to a village 20 km distant and stops for lunch at a hotel on the way. Shortly before 2 pm a cyclist leaves the hostel to cycle to the same village. He also stops at the hotel and eventually overtakes the hiker as he reaches the village. The accompanying diagram shows this graphically.

(a) How long did the cyclist stop at the hotel?

(b) What was the hiker's speed after lunch? (Give answer to nearest kilometre per hour.)

(c) How far ahead of the cyclist was the hiker when the cyclist resumed his journey from the hotel?

(d) At what time would a motorist travelling at 60 km per hour have to leave the hostel in order to reach the village simultaneously with the hiker and the cyclist?

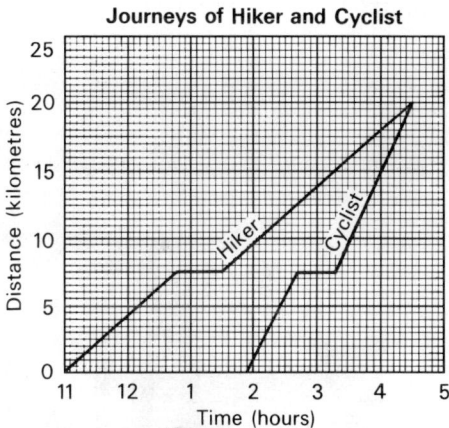

Journeys of Hiker and Cyclist

Distribution of U.K. External Trade
1982 compared with 1973

Exports   Imports

1983 ☐ 60533 ⎫ £'s MILLION   1983 ☐ 65993 ⎫ £'s MILLION
1973 ▨ 12087 ⎭   1973 ▨ 15723 ⎭

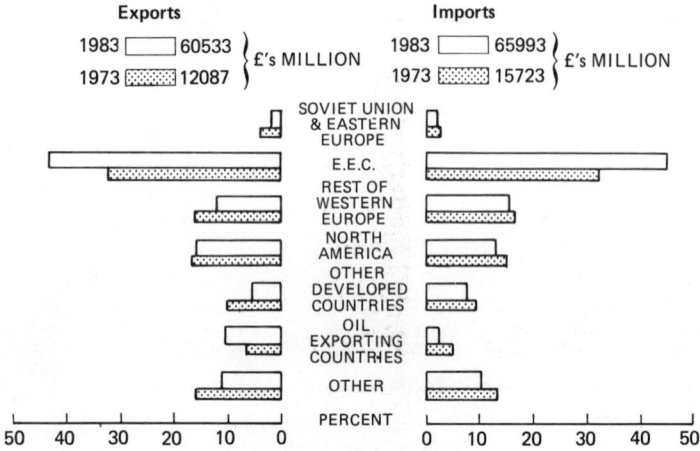

SOVIET UNION & EASTERN EUROPE

E.E.C.

REST OF WESTERN EUROPE

NORTH AMERICA

OTHER DEVELOPED COUNTRIES

OIL EXPORTING COUNTRIES

OTHER

PERCENT

50 40 30 20 10 0   0 10 20 30 40 50

3 (a) To which areas was the percentage of exports greater in 1983 than in 1973?
   (b) From which area was the percentage of imports greater in 1983 than in 1973?
   (c) In 1983 what percentage of U.K. exports went to North America?
   (d) The total value of exports in 1983 was £60 533 million. What was the value of exports to North America?
   (e) What was the difference between the percentage of imports from E.E.C. countries in 1973 and 1983?
   (f) With which area has the U.K. increased exports from 1973 to 1983 while reducing imports?
   (g) The total value of imports in 1983 was £65 993 million. What was the value of imports from the E.E.C.?

Tell your teacher you have finished this unit

UNIT 10 Rates and Taxes

A Wage deductions

Athur Howlett Ltd also require experienced Bottom Cementer.
Eastern Evening News

yes it's a great job

REINDEER KEEPER required immediately. Single person with driving licence preferred.
John O'Groats Journal

HOLIDAYS ARE EXCELLENT 5½ WEEKS INCLUDING BANK HOLIDAYS!

Full time Shop Ass – apply in person.
Evening Post (Nottingham)

Battery shop assistant required; good prospects to right person: permanent: able to take charge.
Southern Evening Echo

When you start work, pay day will be an important day to you. You may be paid once a week (a, weekly **wage**), or once a month (a monthly **salary**), and you will receive a **wage slip**.

Miss Angela Black is an employee (or worker) in a supermarket. Each week she receives a wage slip like this:

DEPARTMENT	EMPLOYEE	WEEK NO.	TAX CODE	
7	A Black	1		
BASIC PAY	COMMISSION	OVERTIME PAY	BONUS	TOTAL GROSS PAY
85.20	—	—	7.50	92.70
NATIONAL INSURANCE	INCOME TAX	COMPANY PENSION	OTHER DEDUCTIONS	TOTAL DEDUCTIONS
8.34	15.45	—	—	23.79
				NET PAY
				68.91

Employees do not receive all the money they have earned.

Money is deducted for **National Insurance** and **Income Tax**.

Some firms have pension funds, club subscriptions or savings schemes and employees pay into these.

A wage slip tells you how much you have earned (**gross pay**),

how much has been taken out of your wage (**deductions**),

how much actual money you have been paid (**net pay**).

TOTAL GROSS PAY	92.70
TOTAL DEDUCTIONS	23.79
NET PAY	68.91

So, Net pay = Gross pay − Deductions

Exercise

1 Calculate the amounts (a) to (d) in the table below:

Gross pay	Total deductions	Net pay
£95.46	£17.82	(a)
£109.80	£20.52	(b)
(c)	£28.02	£116.28
£117.48	(d)	£94.04

Example

Gordon's total gross pay is £107.50 per week.
From this the following deductions are made:

National Insurance £ 9.70
Income Tax £12.90
  Total deductions £22.60

Gordon's net pay = Total gross pay − Total deductions
  = £107.50 − £22.60
  = £84.90

Exercise

2

Jim's gross weekly wage is £115.82. From this the following deductions are made:

Copy and complete:

National Insurance = £10.42
Income Tax = £15.41
Total deductions = £ ▓▓▓

Net pay = Gross pay − Deductions
  = £ ▓▓▓ − £ ▓▓▓
  = £ ▓▓▓

3 Fred is a fireman earning a total weekly wage of £216. His deductions are: Income Tax £45.40, National Insurance £19.44, Union dues £4.50. Find Fred's total deductions and his net weekly wage.

4 From Miss McTavish's monthly pay slip find her net pay.

DEPARTMENT —	EMPLOYEE	WEEK NO. —	TAX CODE	Miss R. McTavish	
BASIC PAY 691.50	COMMISSION —	OVERTIME PAY —	BONUS —	TOTAL GROSS PAY	
NATIONAL INSURANCE 62.24	INCOME TAX 153.90	COMPANY PENSION 30.80	OTHER DEDUCTIONS 3.20	TOTAL DEDUCTIONS	
					NET PAY

5 Angus MacGibbon earns a gross monthly salary of £840. Each month his deductions are:

Income Tax   £168.15
National Insurance   £ 75.60
Sports Club Subscription   £ 1.60

Calculate his total deductions and his net monthly pay.

6 Check whether the following wage slips are correctly calculated. Where you find a mistake work out the correct net pay.

(a)

DEPARTMENT	EMPLOYEE 27	WEEK NO. 3	TAX CODE	Duncan Millar	
BASIC PAY 116.25	COMMISSION —	OVERTIME PAY 48.60	BONUS 14.55	TOTAL GROSS PAY 179.40	
NATIONAL INSURANCE 16.15	INCOME TAX 34.50	COMPANY PENSION —	OTHER DEDUCTIONS 0.90	TOTAL DEDUCTIONS 51.55	
				NET PAY 128.85	

(b)

DEPARTMENT	EMPLOYEE 1	WEEK NO. 10	TAX CODE	Susan Tuck	
BASIC PAY 74.60	COMMISSION —	OVERTIME PAY —	BONUS 7.20	TOTAL GROSS PAY 81.80	
NATIONAL INSURANCE 10.19	INCOME TAX 12.18	COMPANY PENSION —	OTHER DEDUCTIONS —	TOTAL DEDUCTIONS 22.37	
				NET PAY 59.43	

(c)

DEPARTMENT	EMPLOYEE 4	WEEK NO. 5	TAX CODE	Janet Williams	
BASIC PAY 89.40	COMMISSION 55.80	OVERTIME PAY 29.40	BONUS —	TOTAL GROSS PAY 173.60	
NATIONAL INSURANCE 15.71	INCOME TAX 33.40	COMPANY PENSION —	OTHER DEDUCTIONS 2.99	TOTAL DEDUCTIONS 50.10	
				NET PAY 123.50	

B Gross pay

In some jobs workers receive more than the basic wage through **overtime** or **bonus**.

Exercise

1

	TIME SHEET			
Work done for R.D. Workenough Ltd				
Name J. Right Workman I Week ending 5·4·77				
DAY	DESCRIPTION OF WORK	HOURS	RATE	AMOUNT
MON	Installation of inspection tank	8	£3·45	27·60
	overtime	2	£4·60	9·20
TUE	Repair fault in inspection tank	8	£3·45	
	overtime	1	£4·60	
WED	Repairs to Battle St. High School	8	£3·45	
	overtime	3	£4·60	
THUR	Extend wall at Battle St. High School	8	£3·45	
	overtime	2	£4·60	
FRI	Demolition of Project B.	8	£3·45	
	overtime	—		
SAT	Complete demolition work on Project B			
	overtime	5	£5·18	
			GROSS PAY	£

Find J. Nisbet's gross weekly wage by copying and completing this column.

2

NAME Norman Gallagher				
WEEK ENDING 25·1·77				
			Ordinary Time	Over-Time
MON	0800	1200	4	
	1300	1700	4	
TUE	0800	1200	4	
	1300	1700	4	
WED	0800	1200	4	
	1300	1900	4	2
THUR	0800	1200	4	
	1300	1800	4	1
FRI	0800	1200	4	
	1300	1700	4	
SAT	0800	1200		4
Total Hours			40	7

	HOURS	RATE	Amount
ORDINARY TIME	40	2·82	
OVERTIME	7	3·76	
GROSS WAGES.........			
DEDUCTIONS:			
NAT. INSUR. £ 12·49			
INCOME TAX £ 22·30			
NET WAGE			£

Find Norman's *net* weekly wage if he works 40 hours at the basic rate of £2.82 per hour and 7 hours at £3.76 per hour.

His deductions are shown on his card.

Continue with Section C

C Overtime

John Ferguson's basic wage is £3.46 per hour. From Monday to Thursday he works from 8 am till 12 noon and from 1 pm till 5 pm. On Friday he finishes at 4 pm. That is, he works 8 hours a day for 4 days a week and 7 hours on Friday – a basic 39 hour week.

So, his **basic weekly wage** is
39 × £3.46 = £134.94.

If he works **overtime** during the week, that is, if he works after 5 pm (or 4 pm on Friday), he is paid $1\frac{1}{2}$ times the basic hourly rate.

We say he is paid **time and a half**.

So, the overtime rate from Monday to Friday is

$$1\tfrac{1}{2} \times £3.46 = £5.19 \text{ per hour.}$$

If he works overtime at the weekend he is paid even more.

For working on Saturday or Sunday he is paid twice the basic hourly rate.

We say he is paid **double time**.

So, the overtime rate for Saturday and Sunday is

$$2 \times £3.46 = £6.92 \text{ per hour.}$$

Example

Find his gross wage in a week when he works his basic 39 hours, together with 1 hour overtime on Monday, 2 hours overtime on Wednesday, and 4 hours overtime on Saturday.

Basic earnings:	39 × £3.46 =	£134.94
Overtime: 3 hours on time and a half =	3 × £5.19 =	£ 15.57
4 hours on double time =	4 × £6.92 =	£ 27.68
John Ferguson's gross wage		£178.19

Exercise

1 A cook is paid at a basic rate of £2.24 per hour. One week she works 6 hours overtime where 4 hours are paid at time and a half and 2 hours are paid at double time.

   (a) How much does she earn in overtime that week?
   (b) If she works a 37 hour week at the basic rate, what is her gross wage for that week?

2   For working overtime, Douglas Gillam is paid:

Monday–Friday: time and a half
Saturday/Sunday: double time

His basic rate is £2.80 per hour for a 39 hour week.

In a certain week he works the following overtime:

Monday   17.00–19.00 hours
Tuesday   17.00–18.00 hours
Wednesday   17.00–20.00 hours
Saturday   8.00–12.00 hours

Find his gross wage for the week.

3   Steve delivers parcels by motorbike. His basic hourly rate for a 38 hour week is £2.40.

Overtime is paid for at time and a third.

What is his gross wage in a week when he works 5 hours overtime?

4   Bert is a garage mechanic. His basic hourly rate for a 39 hour week is £2.76. Any over-time he works is paid at time and a third.

What is his gross wage in a week when he works 10 hours overtime and is paid a bonus of £11.20?

5   Katie works in a hotel at weekends. The basic rate is £1.86 per hour. Evening work (after 18.00 hours) is paid at time and a half.

If Katie works on Saturdays from 10.00 hours to 22.00 hours and on Sundays from 16.00 hours to 20.00 hours, what are her weekend earnings?

Continue with Section D

D Commission

In some jobs workers receive more than the basic wage through **commission**.

Example

Mr Dixon is a used car salesman. He receives a basic monthly salary of £512 together with 3% commission on all sales. Find his gross earnings in a month in which his sales amount to £24 600.

			Working
Basic salary		$=$£ 512	
Commission: 3% of £24 600		$=$£ 738	£
Mr Dixon's gross earnings		$=$£1250	1% of £24 600 $=$ 246
			3% of £24 600 $=$ 738

Exercise

1   A salesman earns a basic salary of £520 per month, plus 5% commission on his sales. Find his gross income in a month in which he sells £5700 worth of goods.

2   Calculate the gross income per month of each of the following:

	Basic salary per month	Commission on sales	Sales
(a)	£680	6%	£5400
(b)	£400	9%	£6800
(c)	£880	4%	£15788

3   The flowchart below shows the method of calculating the monthly salary of a salesman who is paid a basic salary plus a commission on sales.

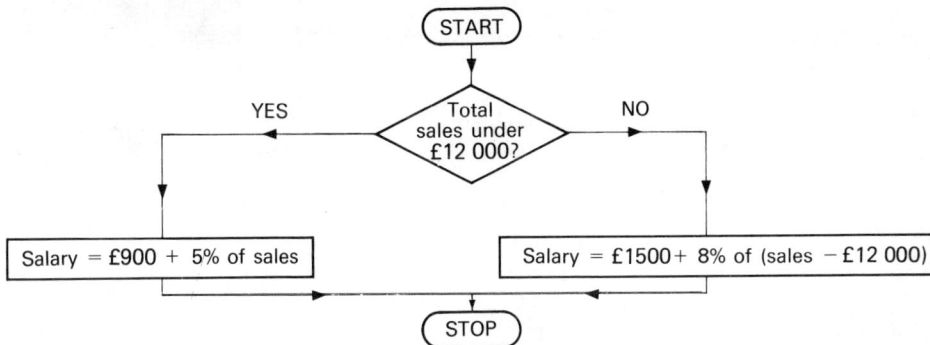

What are the salaries payable to salesmen with total sales of (a) £10 800, (b) £16 800?

Continue with Section E

E National Insurance contributions

Pensions

Sickness Benefits

National Insurance
Payments
help to finance:

Injury
Benefits

National Health Service

Unemployment Benefits

Each week a worker and his employer pay a contribution to the National Insurance Fund.

The amount of the contribution depends upon the worker's gross wage.

Weekly National Insurance contribution rates are shown in the table opposite.

Example

If a worker earns a gross wage of £162.00 his contribution is £14.60 and his employer's contribution is £16.96. Every week the worker's share is deducted from his gross wage along with the other deductions before he is paid.

Exercise

1 (a) Use the table to find how much an employee pays in National Insurance if his gross pay is

(i) £134.50   (ii) £120   (iii) £146.50   (iv) £140.40   (v) £128.60

(b) How much does the employer pay for each of these people? (Subtract the employee's contribution from the total of employee's and employer's contributions.)

2 In a family, the father earns £163 a week, the son earns £138 a week, and the daughter earns £105 a week. Find the *total* amount this family pays in National Insurance contributions in a week.

3 An Aberdeen man earns a gross weekly wage of £107.50. Find the total amount he pays in National Insurance contributions in a year, that is 52 weekly contributions.

4 An employer pays his share of National Insurance for 3 employees whose gross weekly wages are (a) £115.50, (b) £120, and (c) £139.50. How much does the employer pay altogether?

NATIONAL INSURANCE CONTRIBUTION TABLES
Standard-rate Contribution

If the exact gross pay figure is not shown in the table, use the next smaller figure shown.

Gross Pay	Total of Employee's and Employer's Contrib'ns	Employee's Contrib'n	Gross Pay	Total of Employee's and Employer's Contrib'ns	Employee's Contrib'n	Gross Pay	Total of Employee's and Employer's Contrib'ns	Employee's Contrib'n
£	£	£	£	£	£	£	£	£
103.50	20.18	9.34	123.50	24.07	11.14	143.50	27.96	12.94
104.00	20.27	9.38	124.00	24.16	11.18	144.00	28.05	12.98
104.50	20.38	9.43	124.50	24.27	11.23	144.50	28.16	13.03
105.00	20.47	9.47	125.00	24.36	11.27	145.00	28.25	13.07
105.50	20.57	9.52	125.50	24.46	11.32	145.50	28.35	13.12
106.00	20.66	9.56	126.00	24.55	11.36	146.00	28.44	13.16
106.50	20.77	9.61	126.50	24.66	11.41	146.50	28.55	13.21
107.00	20.86	9.65	127.00	24.75	11.45	147.00	28.64	13.25
107.50	20.96	9.70	127.50	24.85	11.50	147.50	28.74	13.30
108.00	21.05	9.74	128.00	24.94	11.54	148.00	28.83	13.34
108.50	21.15	9.79	128.50	25.04	11.59	148.50	28.93	13.39
109.00	21.25	9.83	129.00	25.14	11.63	149.00	29.03	13.43
109.50	21.35	9.88	129.50	25.24	11.68	149.50	29.13	13.48
110.00	21.44	9.92	130.00	25.33	11.72	150.00	29.22	13.52
110.50	21.54	9.97	130.50	25.43	11.77	150.50	29.32	13.57
111.00	21.64	10.01	131.00	25.53	11.81	151.00	29.42	13.61
111.50	21.74	10.06	131.50	25.63	11.86	151.50	29.52	13.66
112.00	21.83	10.10	132.00	25.72	11.90	152.00	29.61	13.70
112.50	21.93	10.15	132.50	25.82	11.95	152.50	29.71	13.75
113.00	22.02	10.19	133.00	25.91	11.99	153.00	29.80	13.79
113.50	22.13	10.24	133.50	26.02	12.04	153.50	29.91	13.84
114.00	22.22	10.28	134.00	26.11	12.08	154.00	30.00	13.88
114.50	22.32	10.33	134.50	26.21	12.13	154.50	30.10	13.93
115.00	22.41	10.37	135.00	26.30	12.17	155.00	30.19	13.97
115.50	22.52	10.42	135.50	26.41	12.22	155.50	30.30	14.02
116.00	22.61	10.46	136.00	26.50	12.26	156.00	30.39	14.06
116.50	22.71	10.51	136.50	26.60	12.31	156.50	30.49	14.11
117.00	22.80	10.55	137.00	26.69	12.35	157.00	30.58	14.15
117.50	22.90	10.60	137.50	26.79	12.40	157.50	30.68	14.20
118.00	23.00	10.64	138.00	26.89	12.44	158.00	30.78	14.24
118.50	23.10	10.69	138.50	26.99	12.49	158.50	30.88	14.29
119.00	23.19	10.73	139.00	27.08	12.53	159.00	30.97	14.33
119.50	23.29	10.78	139.50	27.18	12.58	159.50	31.07	14.38
120.00	23.39	10.82	140.00	27.28	12.62	160.00	31.17	14.42
120.50	23.49	10.87	140.50	27.38	12.67	160.50	31.27	14.47
121.00	23.58	10.91	141.00	27.47	12.71	161.00	31.36	14.51
121.50	23.68	10.96	141.50	27.57	12.76	161.50	31.46	14.56
122.00	23.78	11.00	142.00	27.67	12.80	162.00	31.56	14.60
122.50	23.88	11.05	142.50	27.77	12.85	162.50	31.66	14.65
123.00	23.97	11.09	143.00	27.86	12.89	163.00	31.75	14.69

Continue with Section F

F Tax allowances

The Government requires large sums of money to meet the needs of the population for housing, education, defence, road building, and so on.

The main source of money for these purposes is **Income Tax**.

Income Tax is a tax based on the amount a person earns in a year. Certain amounts of a person's income are NOT taxed. These are his **allowances**.

Allowances

From time to time the Government changes the allowances. We will take them as:

(a) Single person's allowance £2200
    Married man's allowance..................................... £3450

(b) Wife's earned income allowance £2200
    Additional personal allowance (given if you have at least one child and either your wife is disabled or you receive only the single person's allowance).. £1200

(c) Dependent relative allowance £100

Example

A man and his wife are both working. Find their total allowances.

Married man's allowance	=£3450
Wife's earned income allowance	=£2200
Total allowances	=£5650

Exercise

Find the *total allowances* for each of the following:

1 A married man with two children, whose wife is disabled.

2 A single man who maintains his aged uncle.

3 A single woman with a child.

Continue with Section G

G Taxable income

Most people have Income Tax deducted from their pay by their employer who then pays the tax to the Government. This method of paying Income Tax is called **PAYE**, which stands for **Pay As You Earn**.

The tax office sends the worker (and his employer) a notice of his **code number**. This is a number that tells them the value of the allowances he is entitled to before he pays Income Tax. His gross income less his allowances is called his **taxable income**.

Example

A single man has a salary of £8955 a year. He receives a dependent relative allowance for his mother whom he supports. Find (a) his total allowances and (b) his taxable income.

(a) Single person's allowance $= £2200$
    Dependent relative allowance $= \underline{£100}$
                Total allowances $= £2300$

(b) Taxable income $=$ Gross income $-$ allowances
$$= £8955 \qquad - £2300$$
$$= £6655$$

Exercise

1 Fred works in an office and has a gross annual income of £11 520. He is a widower with one child aged 5. Find (a) his total allowances and (b) his taxable income.

2 (a) John is a single man earning a gross weekly wage of £110. He supports his elderly mother. Find his taxable income for the year.

(b) John decides to get married. Find his taxable income as a married man. (His mother continues to live with them.)

3 Peter and his wife Ann are both working and their combined gross annual income is £15 660. Find their taxable income.

Continue with Section H

H | Income tax – basic rate

When the taxable income is known, the amount of Tax to be paid can be calculated.
The **BASIC RATE** of tax is 30% of the taxable income.

People with high taxable incomes pay higher rates of tax, but at this stage we shall only deal with the BASIC RATE.

Example

A man's taxable income is £1923. How much tax does he pay?

$$1\% \text{ of } £1923 = £19.23$$
$$\text{So } 30\% \text{ of } £1923 = £19.23 \times 30 = £576.90$$

He pays £576.90 in Income Tax.

Exercise

Find the tax payable on the following taxable incomes:

1   £8220   **2**   £3900   **3**   £2550   **4**   £11 775   **5**   £5673

Example

A single man's gross annual income is £8560. How much tax does he pay?

$$\text{Taxable income} = \text{Gross income} - \text{Allowances}$$
$$= £8560 - £2200$$
$$= £6360$$
$$1\% \text{ of } £6360 = £63.60$$
$$30\% \text{ of } £6360 = £63.60 \times 30 = £1908.00$$

Exercise

6   Find out how much a single man will pay in Income Tax if his gross annual income is:

(a) £6672   (b) £8812   (c) £7720   (d) £16 040

7   Find out how much a married man will pay in Income Tax if his gross annual income is:

(a) £7124   (b) £9640   (c) £13 352   (d) £15 520

Continue with Section I

Income tax problems

Example

Mr Brown has one child, and his wife is disabled. His salary is £18 640. Find (a) his total allowances, (b) his taxable income, and (c) his Income Tax for a year.

(a) *Allowances*

Married man's allowance $= £3450$

Additional personal allowance $= £1200$

Total allowances $= £4650$

(b) Taxable income $=$ Gross income $-$ Allowances
$$= £18\ 640 - £4650$$
$$= £13\ 990$$

(c) Tax $= 30\%$ of £13 990

1% of £13 990 $= £139.90$

30% of £13 990 $= £139.90 \times 30 = £4197.00$

He pays £4197.00 in Income Tax.

Exercise

1 A man and his wife are both working, and their combined income is £16 200. Calculate (a) their total allowances, (b) their taxable income, and (c) the Income Tax paid in a year.

In each of the next two questions calculate:

(a) the gross income for a year, (b) the total allowances, (c) the taxable income, and (d) the Income Tax for a year.

2 Mrs Smith is a widow with two children. She works as a secretary and earns £195 per week.

3
Bert is a painter earning £198 per week. He is married and he supports his wife's mother.

Continue with Section J

J | Progress check

Exercise

1   Calculate the amounts (a) to (d) in the table below.

Gross pay	Total deductions	Net pay
£225	£41.04	(a)
£130.20	£27.48	(b)
(c)	£26.01	£79.47
£112.44	(d)	£98.67

2   Isobel McSteven earns a gross monthly salary of £1170. Each month her deductions are:

Income tax £296.00
National Insurance . . . . £ 22.50
Athletic Association . . . . £ 2.46

Find (a) her total deductions and (b) her net pay.

3   Jean is a typist, her basic rate of pay is £3.60 per hour for a 35 hour week. She is paid time and a half for overtime. Find her gross weekly wage for a week in which she works 6 hours overtime.

4   A salesman earns a basic salary of £960 per month, plus 5% commission on his sales. Find his gross income in a month in which he sells £7840 worth of goods.

5   A married man earns £14 920 per year. He has 3 children and his wife is disabled.

Find (a) his total allowances, (b) his taxable income, and (c) his Income Tax for a year.

Ask your teacher what to do next

K Higher tax rates

The tax problems so far have only involved Basic Rate tax. The rate of tax is higher for higher taxable incomes. This may be shown in a table or using a flowchart.

Taxable income	Rate of tax
On the first £14 000	30% – this is called the **BASIC RATE**
On the next £3 000	40% – this is at a **HIGHER RATE**

and the rate increases as the taxable income increases

Income tax flowchart (taxable incomes up to £17 000

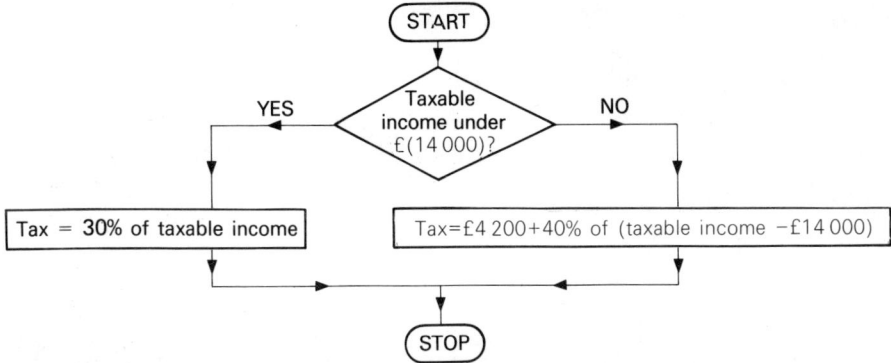

Example

A married man has a salary of £22 350. He maintains his mother. Find (a) his total allowances, (b) his taxable income, and (c) his Income Tax for a year.

(a) Married man's allowance $=£3450$
      Dependent relative allowance $=£\ 100$
               Total allowances $=£3550$

(b) Taxable income $=$ Gross income $-$ Allowances
                    $=£22\ 350-£3550$
                    $=£18\ 800$

(c) BASIC RATE tax $=30\%$ of £14 000 $=£4200$
      HIGHER RATE tax on £(18 800 $-$ 14 000)
         that is, on £4800, is 40% of £4800 $=£1920$
                    Total Income Tax $=£6120$

Exercise

In each question, find the Income Tax for a year.

1   Hughie is a postman. He is not married and earns £7200 per annum.

2   A single man whose gross income is £22 826 maintains his aged mother.

3   Dr Spick earns £24 000 per annum. He is married and he maintains an elderly uncle.

Continue with Section M

L VAT (Valued Added Tax)

The prices of many things we buy (for example clothing, electrical goods, and meals in restaurants) include a tax which is paid to the Government. This is called **Valued Added Tax** (**VAT** for short).

The rate can be varied, but we will work here with a rate of 15%.

In some cases prices are quoted without VAT, and VAT must be added to find the total cost.

Example

Some hotels quote prices which do not include VAT. In this case, what is the total bill when the listed charge is £42?

$$1\% \text{ of } £42 = £0.42$$
$$15\% \text{ of } £42 = £0.42 \times 15 = £6.30$$
$$\text{So total bill} = £42 + £6.30 = £48.30$$

Exercise

1   Find the total bill for each item listed below, when VAT at 15% is added.

(a) a car battery listed as £24   (b) a hotel bill of £57.40
(c) a car repair bill of £97.60   (d) a house repair bill of £578

Example

An Educational authority can claim back **VAT** from the Government on equipment it purchases. A computer is priced at £391, including **VAT**. How much of this price is due to **VAT**, and what is the cost when **VAT** is reclaimed?

The price including VAT = basic price + 15% of basic price
$$= 100\% \text{ of basic price} + 15\% \text{ of basic price}$$
$$= 115\% \text{ of basic price}$$

115% of basic price   = £391
1% of basic price   = £391 ÷ 115 = £3.40
15% of basic price   = £3.40 × 15 = £51.

So VAT is £51 and cost after VAT is reclaimed is £340.

Exercise

2   Find the cost, with VAT deducted, for each item below:

(a) a calculator priced at £9.20   (b) a typewriter at £402.50
(c) an electric keyboard at £221.95   (d) a photocopier at £1392.65

Continue with Section M

M Local government rates

In addition to paying money to the Government, the citizen is required to finance his own local authority which requires money for swimming baths, libraries, schools, parks, street–lighting, and so on.

The money required for these items is raised by **the rates**.

This rating tax is a tax on buildings.

A local valuation officer assesses the **rateable value** of a building by taking into account its locality, its floor space, and its amenities.

The amount paid to the local council by the occupier depends on the **rateable value** of his property.

Attractive house comprising lounge/ dining room, fitted kitchen, cloakroom, 4 bedrooms, bathroom, garage, garden.

Rateable value £468

Each year the council decides how much the occupier is taxed for every £1 of rateable value of his property.

This amount is called in **the rate in the £**.

If the owner has to pay 85p for every £1 of rateable value, then we say that the rate is **85p in the £**.

Example

The rateable value of the house in the illustration is £468. The local rating tax is 85p in the £. How much does the occupier pay in rates?

For £1 of rateable value the occupier pays 85p
So, for £468 of rateable value the occupier pays $468 \times 85p$
$$= 39780p$$
$$= £397.80$$

The occupier pays £397.80 in rates.

Exercise

1 Find how much the rates are for each of the following properties. The local rate in the £ is given below each.

(a) Bungalow: rateable value £320
Rate: 80p in the £

(b) Flat: rateable value £260
Rate: £1.10 in the £

(c) House: rateable value £408
Rate: 95p in the £

2 Shops and offices in the centre of a city generally have a high rateable value. Find the rates payable on office premises with a rateable value of £4130, if the city rates are £1.26 in the £.

3

The rateable value of this house is £476. One year the city rate was £1.13 in the £, and the following year the rate was raised to £1.24 in the £. What was the increase in the house-holder's rates bill?

4 Mr McLaughlin lived in a city where the rate was £1.40 in the £. His house had a rateable value of £520. When he retired he moved to a country cottage whose rateable value was £270 and where the rates were 86p in the £. How much less did he pay in rates?

5 In 1984 the rateable value of a house was £300 and the local rate was £1.25 in the £. After revaluation in 1985, the rateable value of the house was increased to £425 and the local rate was reduced to £0.90 in the £. How much more or less had the occupier to pay in rates?

Continue with Section N

N Calculation of rate per £

Each year the Regional and District Authorities estimate how much they will need to spend in the following year. The table shows estimates for 1984–85

NET ESTIMATED EXPENDITURE payable out of the REGIONAL and DISTRICT RATES for 1984/85

BORDERS REGION	Estimated Net Expenditure		BERWICKSHIRE DISTRICT	Estimated Net Expenditure	
Education	24 456 000		Central Administration	71 000	
Roads and Transportation	6 420 000		Environmental Services		
Social Work	4 604 000		(Cleansing, Parks,		
Drainage	2 297 000		Burial Grounds and		
Public Water	558 000		Other Services)	693 000	
Planning and Development	846 000		Leisure and Recreation		
Libraries	786 000		(Swimming Pools,		
Other Services	2 671 000		Halls and Museums)	162 000	
Police	2 848 000		Tourism	20 000	
Fire	1 295 000		Miscellaneous Services (Grants		
River Purification	192 000		to Village Halls,		
Provision for Inflation	2 100 000		Sports Clubs,		
			Community Councils		
TOTAL TO BE FINANCED	£49 073 000		and Other Organisations.		
LESS			District Court and		
			Licensing Board)	35 000	
Rate Support Grant 32 709 000			Housing (Rebates, Grants		
paid			and Loans)	128 000	
Balance in hand at 1 426 000					
1 April			TOTAL TO BE FINANCED	£1 109 000	
	34 135 000		LESS		
			Rate Support Grant	597 000	
Amount to be raised by	£14 938 000	86p in the £	Amount to be raised by	£ 512 000	17p in the £
REGIONAL RATE			DISTRICT RATE		

We see that £14 938 000 had to be raised by the Borders Region Rate and £512 000 by the Berwickshire District Rate.

The **total** rateable value of *all* the buildings in the Region and District is known and so the amount of the rating tax (or rate in the £) can be calculated.

Example

The total rateable value of a town is £12 000 000 and the local authority's estimated expenditure (that is, spending) which has to be raised from the rates is £9 000 000. Calculate the rate in the £ which the authority must set.

A rateable value of £12 000 000 has to provide £9 000 000.

So a rateable value of £1 has to provide £9 000 000 ÷ 12 000 000

$$= £\frac{9\,000\,000}{12\,000\,000}$$

$$= £\frac{9}{12} = £\frac{3}{4} = £0.75$$

That is, for every £1 of rateable value, a householder pays the council 75p.

The rate is 75p in the £.

Exercise

The rateable values of various towns and their estimated expenditures is given below. Calculate the rate in the £ which must be set to meet this spending.

	Total rateable value (£)	Expenditure (£)
1	8 000 000	5 600 000
2	1 000 000	850 000
3	620 000	744 000
4	850 000	1 479 000

Continue with Section O

O Rate per £ involving a surplus

In the last section the rate per £ was found to be an exact value. In practice this rarely happens. The authority has to adjust the rate per £ so that enough money is gathered to meet the expenditure.

Example

The total rateable value of a town is £9 000 000 and its proposed expenditure amounts to £11 000 000. Find the rate in the £ which the council must set to meet this expenditure.

A rateable value of £9 000 000 has to provide £11 000 000.
So the rateable value of £1 has to provide £11 000 000 ÷ 9 000 000

$$= £\frac{11}{9} = £1.222\ldots$$

If the council set a rate of £1.22 in the £, they would raise £1.22 × 9 000 000 = £10 980 000 which is £20 000 *short* of the total amount needed.

If the council set a rate of £1.23 in the £, they would raise £1.23 × 9 000 000 = £11 070 000 which gives a *surplus* of £70 000.

The council never wish to be *short* of money, so they set the rate at £1.23 in the £, that is, rounded to the *nearest penny above* £1.222 ... .

Example

Calculate the rate in the £ needed when the proposed expenditure is £810 000 000 and the total rateable value is £800 000 000.

$$\text{Rate in the £} = \frac{\text{Proposed expenditure}}{\text{Rateable value}} = £\frac{81}{80} = £1.0125 \approx £1.02 \text{ (rounding up)}$$

Amount raised = 800 000 000 × £1.02 = £816 000 000
Surplus = £816 000 000 − £810 000 000 = £6 000 000.

Exercise

Calculate the rate in the £ needed in the following. Give the rate to the nearest penny above and find the surplus which will be raised.

	Total rateable value	Proposed expenditure
1	£ 7 000 000	£ 6 200 000
2	£110 000 000	£135 000 000
3	£ 2 100 000	£ 2 000 000
4	£360 000 000	£497 000 000

Continue with Section P

P Progress check

1 The rateable value of a house is £510 and the city rate was set at 93p in the £. How much did the householder have to pay in rates?

2 A house had a rateable value of £546 when the city rate was 95p in the £. Next year the rate was increased to £1.25 in the £. How much *more* did the householder have to pay in rates?

3 A region has a rateable value of £2.2 million and an estimated expenditure of £1.9 million. Calculate the rate per £ to be charged and the surplus over £1.9 million which this will produce.

4 Jim is an engineer. He is not married and earns £390 per week and he supports his mother. Find (a) his total allowances, (b) his taxable income for one year, and (c) his Income Tax for one year.

5 A video recorder is listed at £399.05, including VAT at 15%. What would it cost without VAT?

Tell your teacher you have finished this unit

UNIT 11 Index Numbers

A Increasing costs

Meet the Anderson family.

It consists of Mr. and Mrs. Anderson, their daughter Janet and son Ian.

Mrs. Anderson has to keep a careful note of the amount of money she spends on food. The table shows the average amount she spent on the same quantity of food for breakfast over the years 1983 to 1986.

Breakfast	1983	1984	1985	1986
Sausages	80p	85p	97p	108p
Bacon	90p	96p	125p	139p
Eggs	30p	39p	48p	53p
Totals	200p	220p	270p	300p

Notice that the cost of the same quantity of food has increased from year to year. These increases are shown on the graph.

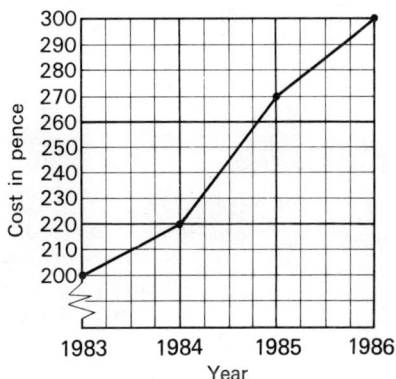

Exercise

1   The average costs for dinners are shown opposite for the years 1983 to 1986.

Copy the table, find the total for each year, and draw a graph of your results on ½ cm squared paper. (Use the same scales as above, but start the vertical scale at 500.)

Dinner	1983	1984	1985	1986
Meat	364p	400p	430p	445p
Potatoes	28p	32p	65p	90p
Vegetables	60p	64p	80p	85p
Pudding	48p	54p	60p	70p
Totals				

2   Here are the average amounts spent on food for tea.

Find the total for each year and draw a graph to illustrate your results.

Tea	1983	1984	1985	1986
Fish	161p	173p	177p	206p
Chips	30p	37p	67p	95p
Bread	37p	45p	50p	63p
Jam	42p	45p	51p	53p
Tea	30p	30p	30p	33p
Totals				

Continue with Section B

B Food index

The average cost of breakfast from 1983 to 1986 is shown below:

Year	1983	1984	1985	1986
Cost in pence	200	220	270	300

In order to measure the changes in cost of the breakfast food we express the cost each year as a percentage of the cost in 1983.

Because 1983 is the 'starting' year we call it the **base year**. The cost in the base year is represented as 100%.

Cost in pence   Percentage

Base year 1983   200 $\longrightarrow$ 100%

$$1 \longrightarrow \frac{100}{200} = \tfrac{1}{2}\%$$

1984   220 $\longrightarrow$ $220 \times \tfrac{1}{2} = 110\%$

1985   270 $\longrightarrow$ $270 \times \tfrac{1}{2} = 135\%$

1986   300 $\longrightarrow$ $300 \times \tfrac{1}{2} = 150\%$

Here is a summary of these results:

Year	1983	1984	1985	1986
Percentage	100%	110%	135%	150%

If we leave out the percent sign (%) we can call these numbers **index numbers**

Index numbers are calculated with reference to a **base year**. The **base year** is always given an index number of 100.

In this example the base year is 1983 and it is usually written as (1983=100).

Mrs. Anderson's 'Breakfast index' is given below:

Year	1983	1984	1985	1986
Cost in pence	200	220	270	300
Breakfast index	100	110	135	150

Exercise

1 The average cost of dinners from 1983 to 1986 from the exercise in Section A is given opposite.

Copy the table and complete the 'Dinner index' taking 1983 as the base year (1983=100).

Year	1983	1984	1985	1986
Cost in pence	500	550	635	690
Dinner index	100	110		

2 The average cost of teas from the same exercise is also given opposite.

Copy the table and complete the 'Tea index' (1983=100).

Year	1983	1984	1985	1986
Cost in pence	300	330	375	450
Tea index				

Continue with Section C

C Beverage index

Example

Mrs Anderson has also kept a check on the price of a small jar of coffee. The table shows the average price of coffee over the years 1983 to 1986.

Year	1983	1984	1985	1986
Price in pence	50	58	68	80

Make a 'Coffee Index' taking 1983 as the base year (1983 = 100).

The cost in the base year is represented by 100.

Price in pence   Index number

Base year 1983   50 $\longrightarrow$ 100

$$1 \longrightarrow \frac{100}{50} = \frac{10}{5} = 2$$

1984   58 $\longrightarrow$ $58 \times 2 = 116$

1985   68 $\longrightarrow$ $68 \times 2 = 136$

1986   80 $\longrightarrow$ $80 \times 2 = 160$

Year	1983	1984	1985	1986
Price in pence	50	58	68	80
Coffee index	100	116	136	160

Exercise

1 Average price of a packet of tea over the years 1983 to 1986 is given opposite.

Copy the table and complete the 'Tea index' taking 1983 as the base year.

Year	1983	1984	1985	1986
Price in pence	50	56	60	68
Tea index	100			

2 Average price of a bottle of orange juice from 1982 to 1985 is given opposite.

Copy the table and complete the 'Orange index' (1982=100).

Year	1982	1983	1984	1985
Price in pence	25	32	40	47
Orange index				

3 Average price of a bottle of milk over the year 1981 to 1984 is as shown.

Copy the table and complete the 'Milk index' (1981=100).

Year	1981	1982	1983	1984
Price in pence	20	22	25	29
Milk index				

Continue with Section D

D Index numbers

The index number system is used for comparing prices, income, or production over a number of years (or weeks or months).

The figure for one year is taken as the base figure (base year = 100) and figures for subsequent years are given as percentages of it.

By this means the rise or fall in prices, incomes, or production can be most readily appreciated.

Example

The average number of students attending a college each evening during the period 1982 to 1985 is shown below:

Year	1982	1983	1984	1985
Number of students	80	88	92	96

Express this in index form (base year, 1982 = 100).

	Number of students		Index number
Base year 1982	80	$\longrightarrow$	100
	1	$\longrightarrow \dfrac{100}{80} = \dfrac{10}{8} =$	$\frac{5}{4}$
1983	88	$\longrightarrow 88 \times \frac{5}{4} =$	110
1984	92	$\longrightarrow 92 \times \frac{5}{4} =$	115
1985	96	$\longrightarrow 96 \times \frac{5}{4} =$	120

Year	1982	1983	1984	1985
Number of students	80	88	92	96
Index number	100	110	115	120

Exercise

In each question, set out your working as above.

1 The population of Britain 1950–1970 was as follows:

Year	1950	1955	1960	1965	1970
Population (in millions)	50	51	52	54	55

Construct a 'Population Index' for these years taking 1950 as the base year (1950 = 100).

2 The quantity of crude oil refined in Britain from 1979 to 1983 is given in the table below:

Year	1979	1980	1981	1982	1983
Crude oil (in million tonnes)	75	81	87	99	111

Express this information in index form (1979 = 100).

3 Coal consumption for the years 1960–1980 was as follows:

Year	1960	1965	1970	1975	1980
Consumption (million tonnes)	200	214	196	170	140

Construct an index for these statistics taking 1960 as the base year.

4 Labour costs in the construction industry from 1979 to 1983 are shown below.

Year	1979	1980	1981	1982	1983
Labour Costs	240	300	360	384	396

Construct an 'Index of Labour Costs' (1979=100) from the data.

Continue with Section E

E Using index numbers

When we know the price of goods in the base year we can estimate the price of the same quantity of goods for any year for which we know the index number.

Because index numbers are designed to give an overall picture of changes, they can only be used as guides to rises and falls in costs.

Example

The cost of electricity to a householder was £160 in 1983. What would you expect him to have paid for the same quantity of electricity in 1985 if the index of electricity prices for that year was 124 (base year, 1983 = 100)?

$$
\begin{array}{llll}
& \text{Index} & & \text{Cost} \\
\textit{Base year 1983} & 100 & \longrightarrow & £160 \\
& 1 & \longrightarrow \dfrac{£160}{100} = & £1.60 \\
\textit{1985} & 124 & \longrightarrow 124 \times £1.60 = & £198.40
\end{array}
$$

Estimated cost of electricity in 1985 = £198.40

Exercise

1 Mrs. Crawford bought what she needed for her Christmas dinner for £34 in 1983 when the food index was 100. How much would you expect her to have paid for the same quantity of goods in 1984 if the food index was 114 (base year, 1983=100)?

2 In 1981 a housewife needed £25 per week to cover the family's food bill. If the food index in 1982 had risen to 116 (1981 = 100) how much would you expect her to need for food in 1982?

3 In 1977 the average wage of a construction worker was £70. If the wage index had risen to 170 by 1981, calculate the new average wage needed to keep pace with general wage rises (1977=100).

4 The weekly cost of electricity to a manufacturer was £250 in 1980. Calculate the estimated cost to the manufacturer of the same quantity of electricity in 1983 and 1984 if the electricity index numbers for these years were (a) 134 in 1983 and (b) 140 in 1984. The base year for this index is 1980 (1980 = 100).

5 A farmer paid £520 for a quantity of fertilizer in 1982. Find what you would expect his cost to be for the same quantity of fertilizer in 1983 and 1984 if the appropriate index numbers for these years were (a) 105 in 1983 and (b) 110 in 1984 (1982=100).

6 In 1980 the cost of a house was £16 000. Find what would be the estimated value of the house in 1981 and 1982 if the index of house prices was (a) 110 in 1981 and (b) 120 in 1982 (1980=100).

Continue with Section F

F Change of base year

When a trade union is discussing a wage increase for its members, it often compares the wages index of its members with the retail price index.
(The **retail price index** measures changes in the cost of goods and services bought by the average household.)

This table shows the Earnings index for 1980–1984 (1980=100)

Year	1980	1981	1982	1983	1984
Earnings index (1980=100)	100	118	131	143	153

The next table shows the retail price index for the same period.

Year	1974		1980	1981	1982	1983	1984
Retail price index (1974=100)	100		250	275	310	325	340

Notice that the base year for the retail price index is 1974 but the base year for the earnings index is 1980.

Because the two sets of index numbers have different base years, they cannot be usefully compared as they are given. We have to make 1980 a base year for the retail price index and calculate new index numbers for the remaining years.

Retail price index

Old index (1974=100)	New index (1980=100)

1980   250 ⟶ 100

$$1 \longrightarrow \frac{100}{250} = \frac{2}{5}$$

1981   275 ⟶ $275 \times \frac{2}{5}$ =110
1982   310 ⟶ $310 \times \frac{2}{5}$ =.....
1983   325 ⟶ ..... =.....
1984   340 ⟶ .... =.....

Year	1980	1981	1982	1983	1984
Earnings index (1980=100)	100	118	131	143	153
Retail price index (1980=100)	100	110			

Exercise

1 Copy and complete the working and table above.

2 Copy each of the following tables of index numbers, make the year indicated the base year and calculate new index numbers for the other years.

(a)
Year	1983	1984	1985
Old index (1979=100)	200	250	280
New index (1983=100)	100		

(b)
Year	1978	1979	1980
Old index (1975=100)	150	180	225
New index (1978=100)			

(c)
Year	1980	1981	1982
Old index (1976=100)	120	138	144
New index (1980=100)			

(d)
Year	1982	1983	1984
Old index (1975=100)	80	72	60
New index (1982=100)			

Continue with Section G

G Comparison of index numbers

The table shows the variation of a wages index in another industry.

Year	1980	1981	1982	1983	1984
Wage index (1978=100)	120	138	144	156	162
Wage index (1980=100)	100				

Exercise

1 (a) Copy the table above. Make 1980 the base year (1980 = 100) and calculate new index numbers for the other years.
   (b) Copy the graph on page 216. Use $\frac{1}{2}$ cm squared paper. Plot the 'Wage index' (1980 = 100), join the points, and label each graph.

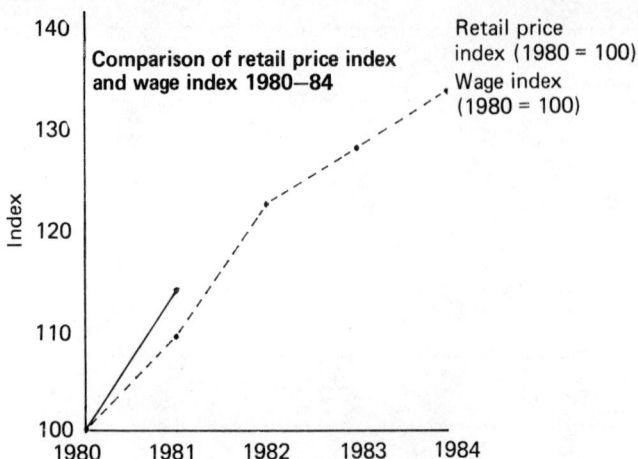

Comparison of retail price index and wage index 1980–84

1 (c) In which year was the wage index above the retail price index?

  (d) In which years was the retail price index the same as the wage index?

  (e) Compared with the base year, what was the percentage increase in wages in 1981?

  (f) In 1980 a girl's weekly wage was £65.

    (i)   Calculate her weekly wage in 1982 using the wage index.
    (ii) What should her wage have been in 1982 to keep up with rising prices? How much extra did she need?

  (g) If a basket of food cost £15 in 1980, how much would you expect the same basket to cost in 1984?.

2 The table shows the steel production index (1980=100) and the aluminium production index (1976=100).

Year	1980	1981	1982	1983
Steel index (1980=100)	100	102	104	106
Aluminium index (1976=100)	120	126	132	138
New aluminium index (1980=100)	100			

Make 1980 the base year for aluminium and calculate new index numbers for the other years to complete the table.
Draw both sets of index numbers on $\frac{1}{2}$ cm squared paper as shown above. (Use the same scales for the axes. Label the vertical scale from 100 to 130, and the horizontal scale, 1980, 1981, 1982, 1983.)

3 The table shows index numbers for the amount of imported sugar and chocolate production.

Year	1981	1982	1983	1984	1985
Sugar index (1981=100)	100	110	98	96	84
Chocolate index (1979=100)	150	156	141	126	123

Make 1981 the base year for chocolate production and calculate new index numbers for the other years.

Show both sets of index numbers on a graph on $\frac{1}{2}$ cm squared paper. (Use the same scales as on opposite page, but start the vertical scale at 80.)

4 The table below shows the petrol price index (1981=100) together with the maintenance and repairs index (1976=100):

Year	1981	1982	1983	1984	1985
Petrol index (1981=100)	100	130	156	178	206
Maintenance index (1976=100)	180	252	270	297	342

(a) Make 1981 the base year for the maintenance index and calculate new index numbers for the other years.
(b) Draw a graph to compare the two sets of index numbers. (Use $\frac{1}{2}$ cm squared paper. Scales: vertical, 1 cm=10 units; horizontal, 2 cm=1 year.)
(c) Name the year when the Petrol index first exceeded the new Maintenance index.
(d) A gallon of petrol cost 97p in 1981, what would it cost in 1985 to the nearest penny?

Continue with Section H

H Progress check

Exercise

1 The table shows the cost in pence of similar quantities of tinned fruit from 1982 to 1985. Calculate the total for each year and express the totals in index form (1982=100).

	1982	1983	1984	1985
Peaches	114	127	130	134
Pears	96	108	110	113
Pineapple	148	·161	164	168
Apricots	142	154	156	160

2 Copy each of the following tables of index numbers and make the year indicated the base year.

Calculate new index numbers for the other years.

Year	1979	1980	1981
Old index (1974 = 100)	400	456	528
New index (1979 = 100)	100		

Year	1981	1982	1983
Old index (1974 = 100)	75	84	93
New index (1981 = 100)			

3 In 1979 the average weekly wage of a waitress was £58. What would you expect her wage to have been in1981 if the wages index was 130 (1979=100)?

4 The used car index was 100 in 1984 and 94 in 1985. What price would you expect to pay in 1985 for a car valued at £4600 in 1984?

5 The table shows index numbers for the amount of energy produced from natural gas and nuclear reactors in Britain.

	1979	1980	1981	1982
Natural gas (1979=100)	100	146	178	224
Nuclear energy (1969=100)	120	150	228	252

(a) Make 1979 the base year for nuclear energy.
(b) Draw a graph on $\frac{1}{2}$ cm squared paper to compare the two sets of index numbers for natural gas and nuclear energy. Name the year when the nuclear energy index exceeded the natural gas index.

Ask your teacher what to do next

▮ Any selected base year

In the following set of index numbers make 1984 a base year and calculate new index numbers for the other years.

Year	1982	1983	1984	1985	1986
Index	110	120	125	130	135

	Old index		New index
1984	125	⟶	100
	1	⟶	$\frac{100}{125} = \frac{4}{5}$
1982	110	⟶	$110 \times \frac{4}{5} = 88$
1983	120	⟶	$120 \times \frac{4}{5} = 96$
Base year 1984	125	⟶	100
1985	130	⟶	$130 \times \frac{4}{5} = 104$
1986	135	⟶	$135 \times \frac{4}{5} = 108$

Year	1982	1983	1984	1985	1986
Old index	110	120	125	130	135
New index	88	96	100	104	108

Exercise

1 Copy each of the following tables of index numbers and make 1981 the base year (1981=100).

Calculate new index numbers for the other years.

(a)

Year	1974	1979	1981	1983
Old index	210	230	250	270
New index			100	

(b)

Year	1980	1981	1983	1985
Old index	144	160	168	184
New index				

2 The index of the price of lemons produced in California U.S.A. was

Year	1979	1980	1981	1982	1983
Lemon index	102	120	111	132	123

Make the base year 1980.

3 House values were assessed in different years as
1973=£4200; 1976=£7000; 1979=£14 000; 1983=£25 200; 1985=£28 000.
Express this in index form (1976=100).

Continue with Section J

J Problems

Example

A trade union was making a claim for an increase in wages for its members. The union officials decided to base the increase on the average wages index. Between January and June 1984 the wages index had risen from 180 to 189. A man's wage was £64 in January 1984; how much should he be paid in June 1984?

	Index		Wage (£)
January	180	$\longrightarrow$	128
	1	$\longrightarrow$	$\dfrac{128}{180}$
June	189	$\longrightarrow$	$189 \times \dfrac{128}{180} = 134.4$

The man should be paid £134.40 in June.

Exercise

1 Between April 1984 and April 1985 the wages index increased from 160 to 176. A man was paid £96 per week in April 1984. How much would he have been paid in April 1985 if his wages increased in the same proportion as the wages index?

2 Between 1983 and 1984 the index of motor car prices increased from 140 to 168. If a motor car cost £2850 in 1983, how much would you have expected it to cost in 1984?

3 The index of coal production decreased from 250 in July 1983 to 225 in December 1983. If 200 million tonnes of coal were produced in July 1983, how many tonnes were produced in December 1983?

4 During a period when the wages index increased from 140 to 152 a man's wage increased from £175 to £187. Is his increase more or less than it would have been if calculated according to the wages index, and what is the difference?

Continue with Section K

K Weighted averages

A building firm paid skilled bricklayers £180 per week and labourers £96 per week. What is the average weekly wage?

If we assume equal numbers of bricklayers and labourers the average weekly wage would be

$$\frac{£180 + £96}{2} = £138.$$

Normally more labourers are employed than bricklayers. Suppose that the firm employed 8 labourers and 4 bricklayers, then the average wage would be:

$$\frac{£96 + £96 + £96 + £96 + £96 + £96 + £96 + £96 + £180 + £180 + £180 + £180}{12}$$

$$= \frac{£1488}{12} = £124$$

Because there are 8 at £96 and 4 at £180 we can write the calculation as:

$$\text{Average weekly wage} = \frac{(8 \times £96) + (4 \times £180)}{12} = \frac{£768 + £720}{12} = \frac{£1488}{12}$$

$$= £124 \text{ (as before)}$$

We say that the average has been 'weighted' by the number of men in each pay group, the 'weights' in this example being 8 and 4.

Because there are twice as many labourers as there are bricklayers we could use the weighting of 2 and 1, like this:

$$\text{Average weekly wage} = \frac{(2 \times £96) + (1 \times £180)}{3} = \frac{£192 + £180}{3} = \frac{£372}{3} = £124$$

$$\uparrow$$

sum of weights

Example

A firm employed 8 men at £20 per week, 4 at £35 per week, and 3 at £40 per week. Calculate the weighted average weekly wage.

Weekly wages (£)	Weights (number of employees)	Wage × weight
120	8	120 × 8 = 960
165	4	165 × 4 = 660
190	3	190 × 3 = 570
	Total weight 15	*Total* 2190

$$\text{Weighted average} = \frac{\text{total of wages} \times \text{weight}}{\text{total weight}} = \frac{2190}{15} = 146$$

The weighted average weekly wage = £146

Exercise

1　Calculate a weighted average wage for each of the following:

(a) Wages (£)	Weight (number of employees)	(b) Wages (£)	Weight (number of employees)	(c) Wages (£)	Weight (number of employees)
60	6	64	10	104	2
111	4	115	8	96	3
180	2	198	2	80	10

2　15 pupils had an average height of 1.6 metres, 20 had an average height of 1.5 metres, and 10 had an average height of 1.35 metres. Calculate the weighted average height of all the pupils.

For certain purposes some examination results are considered to be more important than others. For example, an engineering firm in selecting apprentices considered Mathematics to be four times as important as English, and Physics three times as important as English, so they **weight** the Mathematics, Physics, and English marks by 4, 3, and 1 respectively. This means that the Mathematics mark is multiplied by 4, the Physics mark by 3, and the English mark by 1. The total of these marks is found and this is divided by 8 $(4+3+1 = 8)$ to produce a **weighted average mark**.

Example

The marks obtained by an applicant for a job in the firm are shown below. Calculate the weighted average mark.

Bert

	Marks	Weight	Mark × Weight
Mathematics	50	4	200
Physics	60	3	180
English	78	1	78
Total		8	458

Bert's weighted average mark $= \dfrac{458}{8} = 57.25$

Exercise

3　Calculate Charlie's weighted average mark (Maths 55, Physics 60, English 70). Who will get the job, Bert or Charlie?

4　The examination marks obtained by three pupils applying for a banking post are shown below:

			Pupil	
Subject	Weight	A	B	C
Mathematics	3	47%	31%	74%
English	4	52%	71%	48%
French	1	60%	65%	44%
Geography	2	65%	57%	58%

Calculate the weighted average for each pupil.

Place the three pupils in order of merit according to their weighted averages.

(The data above are weighted to stress the importance of the marks in Mathematics and English.)

5 Using the same examination data as in question 4, calculate the weighted average of each pupil's marks when the following weightings are used in the selection of air hostesses:

Subject	Mathematics	English	French	Geography
Weight	1	2	3	4

Place the pupils in order of merit.

Which two subjects have the greatest influence on the selection of the pupils for this occupation?

Continue with Section L

L Retail price index

The Retail Price Index measures changes in average price levels from month to month of 92 items grouped under the following headings:

	Weight
1. Food	232
2. Alcoholic drink	82
3. Tobacco	46
4. Housing	108
5. Fuel and light	53
6. Durable household goods	70
7. Clothing and footwear	89
8. Transport and vehicles	149
9. Miscellaneous goods	71
10. Services	52
11. Meals bought and consumed outside the home	48
Total of weights	1000

The weights represent the relative importance of the various items in the average family budget and they are revised annually in January.

The 'base year' is also changed when required to keep the index up to date.

The official retail price index is issued monthly by the Department of Employment and its calculation is a complex operation.

We will illustrate one method of calculating a Price Index by considering a limited 'shopping basket' of meat, bread and potatoes. Suppose that in 1984 a survey showed that the average expenditure on meat, bread, and potatoes was:

	1984 prices
Meat	120p
Bread	30p
Potatoes	50p
Total	200p

We calculate the 'weight' for each item by expressing its cost as a percentage of the total cost. (The weight of each item is a measure of its relative importance in the budget.)

	1984 prices		% (or weight)
Meat	120p	⟶	60
Bread	30p	⟶	15
Potatoes	50p	⟶	25
Total	200p	⟶	100

Suppose that in 1985 the cost of similar quantities of these goods had risen and the index number for each was:

meat 110.   bread 125.   potatoes 120   (1984=100)

In order to obtain a single index number which would represent the general price increase we use the 1984 weights to calculate the weighted average of the 1985 index numbers, like this:

	1985 index	weight	index × weight
Meat	110	60	110 × 60 = 6600
Bread	125	15	125 × 15 = 1875
Potatoes	120	25	120 × 25 = 3000
Total weight		100	Total 11475

$$1985 \text{ price index} = \frac{11475}{100} = 114.75 \approx 115$$

The Price Index for 1985=115 (1984=100).

Exercise

1 The table below gives price index numbers in 1981 and the weight relative to the base year (1981=100).

Copy and complete the table and calculate a weighted price index for 1982. (Round answer to nearest whole number.)

	1982 index	weight	index × weight
Butter	102	5	102 × 5 = 510
Bread	98	12	98 × 12 = ▓
Tea	105	3	▓ × ▓ = ▓
	Total weight = ▓		Total = ▓

$$\text{Weighted Price Index for 1982} = \frac{\text{Total '1982 index} \times \text{weight'}}{\text{Total weight}}$$

$$= \frac{▓}{▓} = ▓$$

2 The table lists price index numbers in 1985 and the weights relative to a base year (1982=100). Calculate a weighted index number for 1985. (Round answer to nearest whole number.)

	1985 index	weight
Bacon	110	7
Meat	118	6
Fruit	107	2

3 In 1983 the average expenditure on the following items was

	1983 prices
Bread	40p
Meat	74p
Potatoes	40p
Coffee	12p
Sugar	34p
Total =	200p

(a) Calculate percentage 'weights' for each item for 1983.

(b) In 1985 the price index number for each item was:

bread 150. meat 138. potatoes 150. coffee 167. sugar 183 (1983=100)

Calculate a weighted price index for 1985 (1983=100). (Round answer to the nearest whole number.)

(c) By how much do we expect average expenditure on these items to increase from 1983 to 1985?

4 In 1980 the expenditure patterns in two households were as follows:

	City household	Country household
Food	12%	40%
Tobacco	7%	10%
Housing	31%	20%
Other items	50%	30%

By 1982: Food price index had increased to 120 (1980=100).
Tobacco price index had increased to 110 (1980=100).
Housing cost index had fallen to 90 (1980=100).
All other items had remained stable at 100.

Calculate the weighted average index for each household. Whose expenditure has increased?

Continue with Section M

M Progress check

Exercise

1 Copy and complete each of the following tables making the year indicated the base year:

(a)

Year	1981	1982	1983
Old index (1977=100)	95	125	145
New index (1976 = 100)			

(b)

Year	1983	1984	1985
Old index (1980=100)	180	210	250
New index (1978 = 100)			

2 Between February and August 1983 the index of food prices rose from 180 to 190. In February 1983 a man's wage was £108, what would it need to have been in August 1983 to have kept pace with the increase in food prices?

3 A year ago the wage index was 180 and the industrial production index was 200. They are now 198 and 218 respectively. Calculate the percentage increase in each index and state which has increased more.

4 The table lists price index numbers for agricultural products in 1980, and the weighting for each (1975=100).

Calculate a weighted price index for 1980. (Round answer to nearest whole number.)

	1980 index	weight
Farm crops	94	19
Fatstock	98	32
Livestock	90	41
Fruit and vegetables	92	8

Tell your teacher you have finished this unit

UNIT 12 Estimates and Errors

A Range

600 Pupils held in jail

It is unlikely that exactly 600 pupils were arrested.
The newspaper has given an **approximation** to the actual number.
The actual number is likely to be nearer 600 than 500 or 700.
On a number line the actual number is in this range.

If the number 600 is correct
to the nearest hundred,
the actual number lies between **550** and **650**.

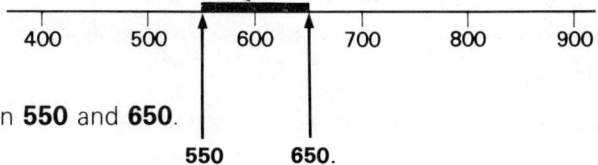

Example

The number 340 is correct to the nearest ten. In what range does the actual number lie?

Draw a number line going up in tens.

Mark the half-way points on either side of 340 and thicken the line between them.

The range is between **335** and **345**.

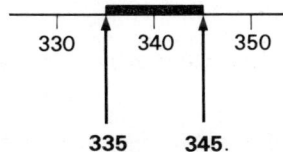

Exercise

For each question draw a number line and find the range in which the actual number lies.

1   720 to the nearest ten
2   5200 to the nearest hundred
3   1800 to the nearest hundred
4   14 000 to the nearest thousand
5   5000 to the nearest thousand
6   400 to the nearest hundred

Continue with Section B

B Rounding: 1 significant figure

For rough calculation purposes it is usual to express numbers correct to **1 significant figure.**

Example

Express 72 correct to 1 significant figure.

From the number line:
  72 lies between 70 and 80 but is nearer 70.
  72 correct to 1 significant figure is 70.

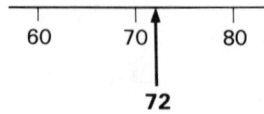

Exercise

Express each of the following correct to 1 significant figure. Set down your working as shown in the example above.

1   43   **2**   81   **3**   89   **4**   19   **5**   66

Example

Express 2342 correct to 1 significant figure.

From the number line:
  2342 lies between 2000 and 3000 but is nearer 2000.
  2342 correct to 1 significant figure is 2000.

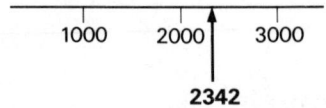

Exercise

Express each of the following correct to 1 significant figure.

6   346   **7**   1753   **8**   66   **9**   740   **10**   96

Example

The following result was obtained by a pupil using a slide rule: $272 \times 19 = 518$. Carry out a rough calculation to see if the result is reasonably accurate.

$272 \times 19 \approx 300 \times 20$ (Each number is correct to
        $= 6000$ 1 significant figure.)

6000 is much greater than 518.
The answer obtained is clearly wrong.

Exercise

Carry out *rough* calculations to see which of the following results are clearly wrong. Set out your working as shown above.

11 $314 \times 29 = 90\,100$ **12** $796 \times 63 = 5020$

13 $1806 \times 96 = 173\,400$ **14** $1340 \times 29 = 39\,000$

15 $1892 \div 23 = 822$ **16** $1284 \div 47 = 2.73$

Continue with Section C

C Rounding: 2, 3 significant figures

Sometimes we are asked to express numbers correct to **two** or **more significant figures.**

Example

Express 268 correct to 2 significant figures.

From the number line:
268 lies between 260 and 270 but is nearer 270.
268 correct to 2 significant figures is 270.

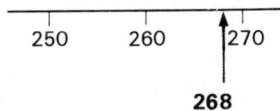

Express 2653 correct to 3 significant figures.

From the number line:
2653 lies between 2650 and 2660 but is nearer 2650.
2653 correct to 3 significant figures is 2650.

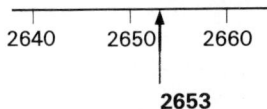

Exercise

Express the following numbers correct to the required number of significant figures. In each case draw number lines and make statements as shown in the examples above.

1 147 (2 sig. figs.) **2** 982 (2 sig. figs.) **3** 2461 (3 sig. figs.)

4 5716 (3 sig. figs.) **5** 1081 (2 sig. figs.) **6** 5190 (2 sig. figs.)

Example

Express 50 249 correct to 2 significant figures.

From the number line:
50 249 lies between 50 000 and 51 000 but is nearer 50 000.
50 249 correct to 2 significant figures is 50 000.

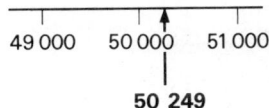

[Note that in this case 5 and 0 are **both** significant figures.]

Exercise

Express the following numbers correct to the required number of significant figures.

7 2033 (2 sig. figs.) **8** 690 357 (3 sig. figs.) **9** 49 862 (2 sig. figs.)

Example

Calculate the exact answer to the following and correct the answer to 2 significant figures:
796×67

$$796 \times 67 = \mathbf{53\,332}$$
$$\approx \mathbf{53\,000} \text{ (to 2 sig. figs.)}$$

Exercise

Calculate the exact answer (using a calculator if you have one) and then give the answer to the required number of significant figures.

10 314×23 (2 sig. figs.) **11** 46×53 (3 sig. figs.)

12 1134×72 (2 sig. figs.) **13** $1036 \div 37$ (1 sig. fig.)

14 $5203 \div 43$ (2 sig. figs.) **15** $3362 \div 82$ (1 sig. fig.)

Continue with Section D

D Rounding involving decimals

Example

Express 17.7 correct to 2 significant figures.

From the number line:
   17.7 lies between 17 and 18 but is nearer 18.
   17.7 correct to 2 significant figures is 18.

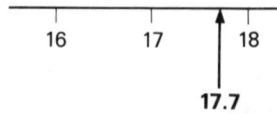

```
16        17    |   18
                |
              17.7
```

Express 9.3 correct to 1 significant figure.

From the number line:
   9.3 lies between 9 and 10 but is nearer 9.
   9.3 correct to 1 significant figure is 9.

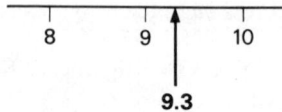

```
8         9  |      10
             |
           9.3
```

Exercise

Express the following numbers correct to the required number of significant figures.

1 18.2 (2 sig. figs.)	**2** 162.8 (3 sig. figs.)	**3** 4.9 (1 sig. fig.)
4 19.6 (2 sig. figs.)	**5** 4.63 (1 sig. fig.)	**6** 18.28 (2 sig. figs.)

Continue with Section E

E Estimates of products

Example

Express 0.84 correct to 1 significant figure.

```
 0.7       0.8  |   0.9
                |
              0.84
```

From the number line:
0.84 lies between 0.8 and 0.9 but is nearer 0.8.
0.84 correct to 1 significant figure is 0.8.

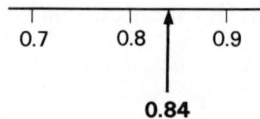

Express 0·075 correct to 1 significant figure.

```
 0.06      0.07  |  0.08
                 |
              0.075
```

From the number line:
   0.075 is half-way between 0.07 and 0.08.
   In this case we go to the *greater* number.
   0.075 correct to 1 significant figure is 0.08.

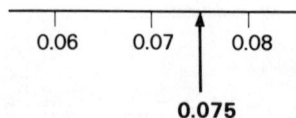

Exercise

Express the following correct to 1 significant figure.

1 0.734	**2** 0.037	**3** 0.15	**4** 0.407

Example

The following result was obtained by a pupil using a slide rule: $0·914 \times 0·73 = 0·667$. Carry out a rough calculation to see if the result obtained is reasonably accurate.

$$0.914 \times 0.72 \approx 0.9 \times 0.7 \quad \left[\begin{array}{l} \text{Each number is correct to} \\ \text{1 significant figure.} \end{array} \right.$$
$$= 0.63$$
$$0.63 \text{ is near to } 0.667$$

The answer obtained is reasonable.

Exercise

Carry out rough calculations to see which of the following results are clearly wrong.

5 $1340 \times 2.9 = 38\,820$

6 $3605 \times 0.84 = 303.5$

7 $477 \times 0.076 = 36.3$

8 $0.914 \times 0.83 = 0.759$

9 $76.4 \times 0.012 = 9.16$

10 $0.845 \times 0.032 = 0.002\,71$

11 $0.562 \div 2.41 = 0.233$

12 $0.187 \div 0.0515 = 0.363$

Continue with Section F

F Rounding decimals

Example

Express 0.150 26 correct to 2 significant figures.

From the number line:
0.150 26 lies between 0.15 and 0.16 but is nearer to 0.15.
0.150 26 correct to 2 significant figures is 0.15.

Express 0.150 26 correct to 3 significant figures.

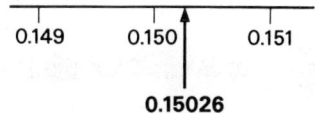

From the number line:
0.150 26 lies between 0.150 and 0.151 but is nearer 0.150.
0.150 26 correct to 3 significant figures is 0.150.

Summary

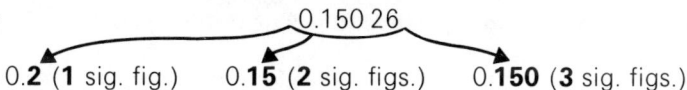

0.150 26

0.**2** (**1** sig. fig.)   0.1**5** (**2** sig. figs.)   0.1**50** (**3** sig. figs.)

Exercise

Express the following numbers correct to the required number of significant figures.

1 0.2637 (1 sig. fig.)

2 0.071 36 (2 sig. figs.)

3 0.8017 (3 sig. figs.)

4 4.263 (1 sig. fig.)

5 4.263 (2 sig. figs.)

6 4.263 (3 sig. figs.)

Example

Calculate the exact answer to the following and correct the answer to 2 significant figures:
76.4×0.12.

$$76.4 \times 0.012 = 0.9168$$
$$\approx 0.92 \quad \text{(to 2 sig. figs.)}$$

Exercise

Calculate the exact answer (using a calculator if you have one) and then give the answer to the required number of significant figures.

7 2.33×0.6 (2 sig. figs.)

8 0.135×0.4 (3 sig. figs.)

9 8.45×3.2 (2 sig. figs.)

10 13.4×2.04 (3 sig. figs.)

11 $4.32 \div 16$ (1 sig. fig.)

12 $6.603 \div 31$ (2 sig. figs.)

Continue with Section G

G Summary

Examples on significant figures can be done using a number line. Once you have understood the number line method, you may wish to use a quicker method. Study the following three examples carefully.

Example

Express 262 865 to 3 significant figures.

262 865
$\approx$ 263 000 (to 3 sig. figs.)

1. Underline the first 3 significant figures.
2. Look closely at the next digit.
3. It is **more** than 5 so the number becomes 263 000.

Express 4926 to 2 significant figures.

4926
$\approx$ 4900 (to 2 sig. figs.)

1. Underline the first 2 significant figures.
2. Look closely at the next digit.
3. It is **less** than 5 so the number stays at 4900.

Express 0.0634 to 2 significant figures.

0.0634
$\approx$ 0.063 (to 2 sig. figs.)

Note that in this example the first two significant figures are the 6 and the 3.

Exercise

Express the following numbers correct to the required number of significant figures.

1 4963 (3 sig. figs.)	**2** 27 199 (2 sig. figs.)	**3** 255 (2 sig. figs.)
4 408 461 (2 sig. figs.)	**5** 690 (1 sig. fig.)	**6** 8.27 (1 sig. fig.)
7 8.27 (2 sig. figs.)	**8** 0.428 (2 sig. figs.)	**9** 0.018 44 (3 sig. figs.)

Continue with Section H

H Estimates using significant figures

Example

Which of the following is nearest in value to $\dfrac{91.6}{0.284}$?

A. 0.3   B. 3   C. 30   D. 300   E. 3000

$$\frac{91.6}{0.284} \approx \frac{90}{0.3}$$

[We start by expressing each number correct to 1 significant figure. This gives a reasonable approximation.]

$$= \frac{900}{3}$$

$$= 300$$

D is the required solution.

Example

Which of the following is nearest in value to 29.3×622?

      A. 1.8   B. 18   C. 180   D. 1800   E. 18 000

$$29.3 \times 622 \approx 30 \times 600$$
$$= 18\,000$$

$$\left[\begin{array}{l} \text{Expressing each number to} \\ \text{1 significant figure.} \end{array} \right]$$

      E is the required solution.

Exercise

1 Which of the following is nearest in value to 69×32?

  A. 21   B. 210   C. 2100   D. 21 000   E. 210 000

2 Which of the following is nearest in value to 82.6×0.66?

  A. 0.056   B. 0.56   C. 5.6   D. 56   E. 560

3 Which of the following is nearest in value to $\dfrac{319}{3.7}$?

  A. 0.8   B. 8   C. 80   D. 800   E. 8000

4 Which of the following is nearest in value to $\dfrac{0.682}{0.0212}$?

  A. 35   B. 3.5   C. 0.35   D. 0.035   E. 0.0035

Example

  Which of the following is nearest in value to $\dfrac{24.5 \times 0.0371}{0.0121}$?

    A. 0.008   B. 0.08   C. 0.8   D. 8   E. 80

$$\frac{24.5 \times 0.0371}{0.0121} \approx \frac{20 \times 0.04}{0.01}$$

$$\left[\begin{array}{l} \text{Express each number correct to} \\ \text{1 significant figure for an approximation.} \end{array} \right]$$

$$= \frac{0.8}{0.01}$$
$$= \frac{80}{1}$$
$$= 80 \qquad \text{E is nearest in value.}$$

Exercise

5 Which of the following is nearest in value to $\dfrac{12.1 \times 0.033}{0.64}$?

  A. 0.005   B. 0.05   C. 0.5   D. 5   E. 50

6 Which of the following is nearest in value to $\dfrac{7.53}{0.414 \times 5.3}$?

  A. 0.004   B. 0.04   C. 0.4   D. 4   E. 40

7 Which of the following is nearest in value to $\dfrac{26.3 \times 4.35}{0.37 \times 12.6}$?

  A. 0.003   B. 0.03   C. 0.3   D. 3   E. 30

Continue with Section I

Error in measurement

No matter how carefully we measure anything we can never know what the correct or true measurement is.

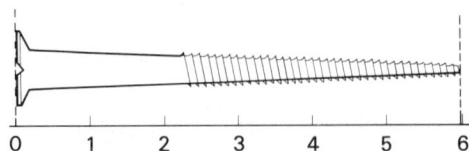

The diagram above shows a screw being measured. We can give its length to the nearest mark on the scale which is 6 units. That is as accurate as we can be using this scale.

If we are given the measurement of an object as 6 units, what we can be sure of is that the true measurement lies between 5.5 units and 6.5 units.

We say that 5.5 units is the **lower limit** and that 6.5 units is the **upper limit**.

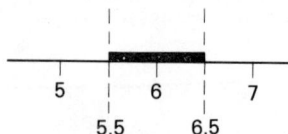

The shortest possible screw falling within this range could not lie more than 0.5 units away from the recorded value of 6 units.

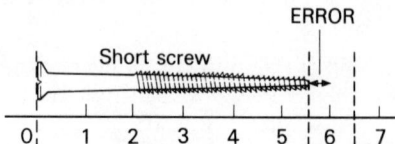

The longest possible screw falling within the 6 units range could not lie more than 0.5 units away from 6.

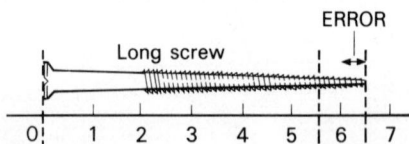

We say that the **greatest possible error** is **0.5 units.**

Example

A measurement is recorded as 19 cm. State (a) the lower and upper limits of the true measurement and (b) the greatest possible error.

(a) 19 cm lies between 18.5 cm and 19.5 cm.
(b) The greatest possible error is 0.5 cm.

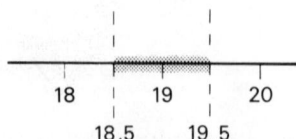

Exercise

For each of the following measurements state (a) the lower and upper limits of the true measurement and (b) the greatest possible error.

1 8 cm   **2** 75 cm   **3** 13 kg   **4** 6 litres

In this diagram each unit has been divided into ten equal parts and so it is possible to give a more accurate reading for the length of the screw.

From the diagram, the length of the screw — to the nearest tenth of a unit — is 5.9 units.

If we are given the measurement of an object as 5.9 units, what we can be sure of is that the true measurement lies between 5.85 units and 5.95 units.

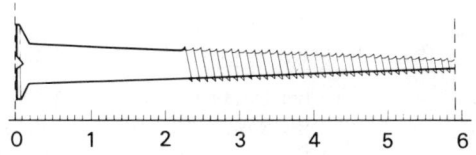

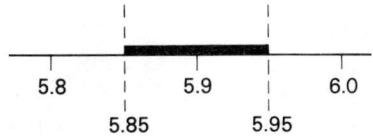

The true measurement could lie not more than 0.05 units from 5.9 units. We say that the **greatest possible error** is 0.05 units.

Example

A measurement is recorded as 17.5 cm. State (a) the lower and upper limits of the true measurement and (b) the greatest possible error.

(a) 17.5 cm lies between 17.45 cm and 17.55 cm.
(b) The greatest possible error is 0.05 cm.

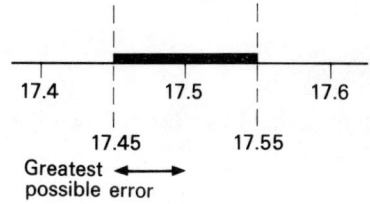

Exercise

For each of the following measurements state (a) the lower and upper limits of the true measurement and (b) the greatest possible error.

5 4.3 g   **6** 15.7 cm   **7** 9.6 kg   **8** 48.3 km

By using a more exact measuring device it is possible to measure to smaller and smaller units.

The screw in the last example, when using a special measuring device, was measured as 5.88 units. What we can be sure of is that the true measurement lies between 5.875 units and 5.885 units.

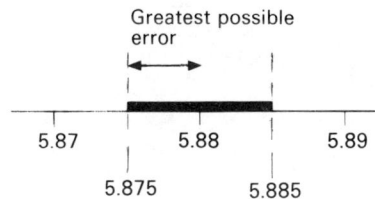

The true measurement could lie not more than 0.005 units from 5.88 units. We say that the **greatest possible error** is 0.005 units.

Example

For each of the following measurements state (a) the lower and upper limits of the true measurement and (b) the greatest possible error.

1 3.87 cm   **2** 17.8 kg   **3** 234 km

1 (a) 3.87 cm lies between 3.865 cm and 3.875 cm.
    (b) The greatest possible error is 0.005 cm.

2 (a) 17.8 kg lies between 17.75 kg and 17.85 kg.
    (b) The greatest possible error is 0.05 kg.

3 (a) 234 km lies between 233.5 km and 234.5 km.
    (b) The greatest possible error is 0.5 km.

Exercise

For each of the following measurements state (a) the lower and upper limits of the true measurement and (b) the greatest possible error.

9 2.24 litres   **10** 18.2 cm   **11** 1.03 cm   **12** 124 m

Continue with Section J

J Error in simple calculations

Example

A man is building a fence between his house and a boundary fence. Allowing for fence posts he reckons he needs four 2.4 metre fence panels.

Find the limits between which the total length of the four panels will lie?

Each panel lies between 2.35 and 2.45 so we have.

Lower limit	Recorded value	Upper limit
2.35	2.4	2.45
×4	×4	×4
9.40	9.6	9.80

So the length of the four panels lies between 9.40 and 9.80.

A lorry is weighed on a weighbridge to measure its load of sand before delivery to a customer. It makes three deliveries with recorded weights of 15 500 kg, 9400 kg, and 12 600 kg. (Each measurement is recorded to the nearest 100 kg.) Find the limits between which the total of the three loads will lie.

Lower limit	Recorded value	Upper limit
15 450	15 500	15 550
9 350	9 400	9 450
12 550	12 600	12 650
37 350	37 500	37 650

So the total of the three loads lies between 37 350 kg and 37 650 kg.

Exercise

1 Find the limits between which the perimeter of a square of side 7.4 cm must lie.

2 The weights of four lorries are recorded as 8300 kg, 7500 kg, 10 600 kg, 9700 kg, (each measurement recorded to the nearest 100 kg). Find the limits between which the total weight will lie.

3 Ten lengths of rail, each 25 m long, are laid end to end. What are the limits of their total length?

4 Find the limits for the total amount of wire required to make a cube of edge 4.5 cm.

5 The number of seeds in a packet is recorded as 50 (to the nearest ten). After filling 10 such packets what are the limits between which the total seeds used must lie.

6 Six tomatoes weigh 63 g, 59 g, 43 g, 66 g, 82 g, and 71 g (to the nearest gram in each case). Find the limits between which the total weight must lie.

Continue with Section K

K Progress check

Exercise

1 The number of spectators at a football match is given as 16 600, correct to the nearest hundred. Draw a number line and find the range in which the actual number lies.

2 Express the following correct to the required number of significant figures.

(a) 376 (1 sig. fig.)   (b) 35 (1 sig. fig.)   (c) 1123 (1 sig. fig.)
(d) 359 (2 sig. figs.)   (e) 1076 (2 sig. figs.)   (f) 405 (2 sig. figs.)
(g) 1056 (3 sig. figs.)   (h) 2392 (3 sig. figs.)   (i) 1075 (3 sig. figs.)

3 Calculate the exact answer, then give the answer to 2 significant figures:

(a) 25×93 (b) $7632 \div 36$

4 Express the following correct to the required number of significant figures.

(a) 0.84 (1 sig. fig.)   (b) 0.037 (1 sig. fig.)   (c) 0.18 (1 sig. fig.)
(d) 0.317 (2 sig. figs.)   (e) 0.0913 (2 sig. figs.)   (f) 28.75 (2 sig. figs.)
(g) 28.75 (3 sig. figs.)   (h) 7.316 (3 sig. figs.)   (i) 0.032 58 (3 sig. figs.)

5 Calculate the exact answer, then give the answer to the required number of significant figures.

(a) 23.5×0.3 (2 sig. figs.)   (b) 14.6×3.03 (3 sig. figs.)

6 Which of the following is nearest in value to $\dfrac{9.3 \times 0.042}{0.56}$?

A. 0.006   B. 0.06   C. 0.6   D. 6   E. 60

7 Which of the following is nearest in value to $\dfrac{27.3 \times 0.63}{8.5 \times 0.023}$?

A. 0.01   B. 0.1   C. 1   D. 10   E. 100

8 State the limits of the measurement and the greatest possible error for:

(a) 3.2 g   (b) 4.34 m   (c) 7 kg   (d) 7.40 cm

9 Find the limits between which the perimeter of a square of side 2.6 cm must lie.

10 Find the limits between which the perimeter of a triangle of side 4.3 cm, 4.7 cm, and 3.9 cm must lie.

Ask your teacher what to do next

L Error in addition

Example

What are the limits of the sum of the measurements 3.2 cm and 4.6 cm, each being given correct to 2 significant figures? State the maximum possible error.

3.2 cm lies between 3.15 cm and 3.25 cm.   4.6 cm lies between 4.55 cm and 4.65 cm.

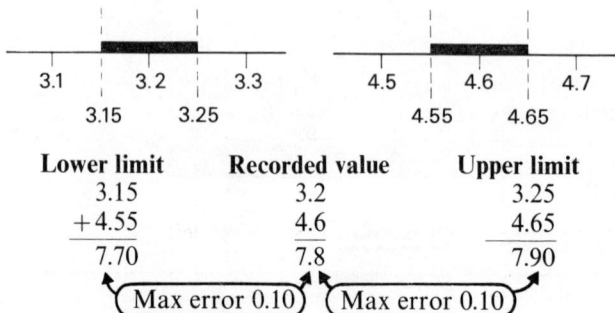

Lower limit	**Recorded value**	**Upper limit**
3.15	3.2	3.25
+4.55	4.6	4.65
7.70	7.8	7.90

Max error 0.10   Max error 0.10

The sum lies between 7.70 cm and 7.90 cm.
The maximum possible error is 0.10 cm.

Example

A rectangle has length 7.5 cm and breadth 5.0 cm. Find the limits of the true measurements of length, breadth, and perimeter, and the maximum possible error of the perimeter measurement.

7.5 cm

5.0 cm   5.0 cm

7.5 cm

Perimeter $= (2 \times \text{length}) + (2 \times \text{breadth})$

7.5 cm lies between 7.45 cm and 7.55 cm.
5.0 cm lies between 4.95 cm and 5.05 cm.

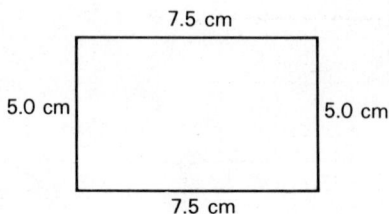

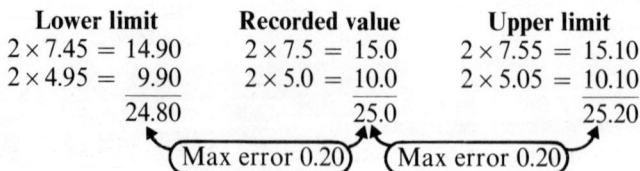

Lower limit	**Recorded value**	**Upper limit**
$2 \times 7.45 = 14.90$	$2 \times 7.5 = 15.0$	$2 \times 7.55 = 15.10$
$2 \times 4.95 = 9.90$	$2 \times 5.0 = 10.0$	$2 \times 5.05 = 10.10$
24.80	25.0	25.20

Max error 0.20   Max error 0.20

The perimeter lies between 24.80 cm and 25.20 cm.
The maximum possible error is 0.20.

Exercise

1   A rectangle has length 12 m and breadth 8 m. What are the limits of the true measurements of length, breadth, and perimeter? What is the greatest possible error in the measurement of the perimeter?

2   What are the limits between which the perimeters of the following metal shapes must lie? Find also the greatest possible error in the perimeter measurement.

   (a) A triangle with sides of lengths 3 cm, 4 cm, and 5 cm.
   (b) A triangle with sides of lengths 3.3 cm, 4.5 cm, and 5.2 cm.

Continue with Section M

M Error in subtraction

Example

The length of a log is 3.5 m. If a piece measuring 0.6 m is sawn off, what is the smallest piece which could be left, and what is the largest?

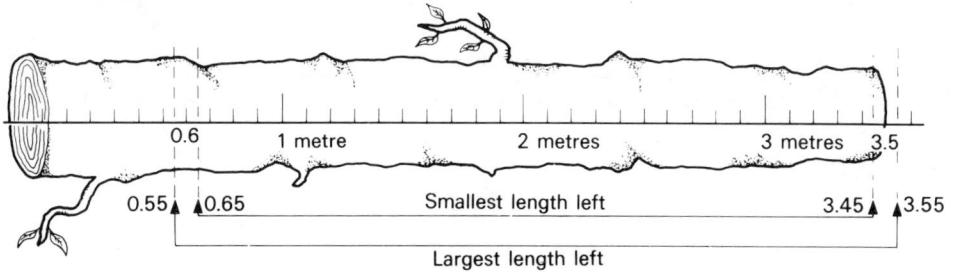

Length of wood = 3.5 m
Length sawn off = 0.6 m
Length left = 2.9 m

But 3.5 m lies between 3.45 m and 3.55 m
and 0.6 m lies between 0.55 m and 0.65 m,
so the smallest piece left = 3.45 m − 0.65 m = 2.80 m

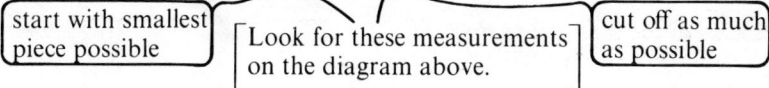

| start with smallest piece possible | Look for these measurements on the diagram above. | cut off as much as possible |

and the largest piece left = 3.55 m − 0.55 m = 3.00 m

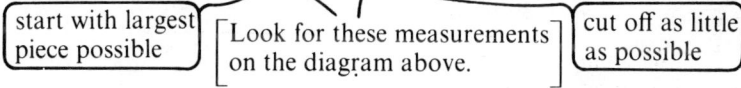

| start with largest piece possible | Look for these measurements on the diagram above. | cut off as little as possible |

Example

A gate 1.3 m wide is fitted between two houses that are 6.4 m apart. What are the limits of the remaining length?

1.3 m lies between 1.25 m and 1.35 m.
6.4 m lies between 6.35 m and 645 m.

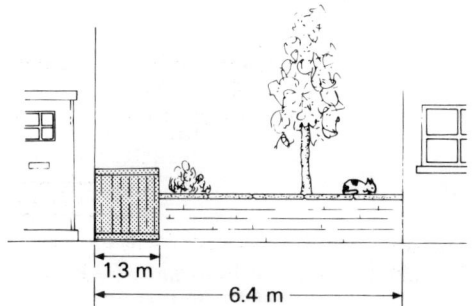

1.3 m
6.4 m

start as small as possible	Lower limit	Recorded value	Upper limit	start as large as possible
	6.35	6.4	6.45	
subtract as much as possible	− 1.35	− 1.3	− 1.25	subtract as little as possible
	5.00	5.1	5.20	

The remaining length lies between 5.00 m and 5.20 m.

Exercise

1   A length of 15 cm is cut from a metal rod 28 cm long.

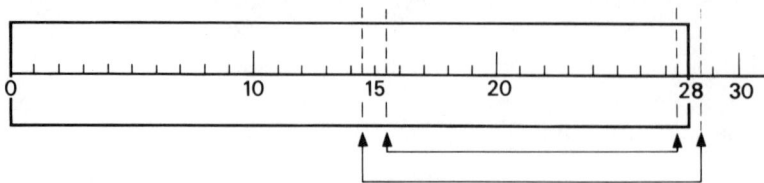

What are the limits of the remaining length?

2   14 kg of flour are removed from a container holding 50 kg. What are the limits of the weight of the remaining flour?

3   (a) A metal strip 10 cm long is cut from a piece 45 cm long. What are the limits of the length left?
(b) If two such strips are cut off, what are the limits of the remaining length?

4   The length of a piece of wood is 2.5 m. If a piece measuring 0.5 m is sawn off, what is the smallest piece which could be left and what is the largest?

Continue with Section N

N Error in calculation of areas

Example

Between what limits does the area of a rectangle of length 5 cm and breadth 4 cm lie?

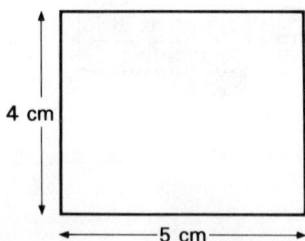

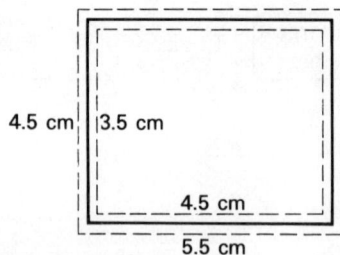

Length = 5 cm   5 cm lies between 4.5 cm and 5.5 cm.
Breadth = 4 cm   4 cm lies between 3.5 cm and 4.5 cm.

Lower limit (smallest possible area)	Recorded value	Upper limit (greatest possible area)
4.5	5	5.5
× 3.5	× 4	× 4.5
2 25	20	2 75
13 50		22 00
15.75		24.75

The area lies between 15.75 cm^2 and 24.75 cm^2.

Exercise

Find the limits between which the areas of the following shapes lie:

1 A rectangle of length 4 m and breadth 3 m.

2 A square with sides of length 3 m.

3 A rectangle of length 8 cm and breadth 2 cm.

Continue with Section O

O Possible values

Example

In the following, which division gives (a) the least value and (b) the greatest value?

A. $36 \div 9$ B. $36 \div 2$ C. $36 \div 6$ D. $36 \div 4$ E. $36 \div 18$

(a) E (*Least* value from $36 \div 18$, which has the *largest* divisor.)
(b) B (*Greatest* value from $36 \div 2$, which has the *smallest* divisor.)

Exercise

In each of the following state which division gives (a) the least value and (b) the greatest value.

1 A. $24 \div 12$ B. $24 \div 3$ C. $24 \div 2$ D. $24 \div 6$ E. $24 \div 8$

2 A. $15 \div 5$ B. $15 \div 50$ C. $15 \div 0.05$ D. $15 \div 0.5$ E. $15 \div 0.005$

3 A. $10 \div 12.3$ B. $10 \div 123$ C. $10 \div 1.23$ D. $10 \div 0.123$ E. $10 \div 0.0123$

4 A. $100 \div 400$ B. $100 \div 0.4$ C. $100 \div 0.04$ D. $100 \div 40$ E. $100 \div 4$

Example

Which division gives (a) the least value and (b) the greatest value?

A. $135 \div 5$ B. $25 \div 5$ C. $50 \div 5$ D. $15 \div 5$ E. $120 \div 5$

(a) D (Least value from $15 \div 5$.)
(b) A (Greatest value from $135 \div 5$.)

Exercise

In each of the following state which division gives (a) the least value and (b) the greatest value.

5 A. $100 \div 4$ B. $16 \div 4$ C. $36 \div 4$ D. $1000 \div 4$ E. $120 \div 4$

6 A. $100 \div 4$ B. $10 \div 4$ C. $0.1 \div 4$ D. $1 \div 4$ E. $0.01 \div 4$

7 A. $12.3 \div 4$ B. $123 \div 4$ C. $1.23 \div 4$ D. $0.123 \div 4$ E. $1230 \div 4$

8 A. $100 \div 12.3$ B. $36 \div 12.3$ C. $199 \div 12.3$ D. $0.69 \div 12.3$ E. $0.2 \div 12.3$

Example

If each of the numbers 3.4 and 2.1 is correct to 2 significant figures, which of the following gives the least possible value of the quotient $3.4 \div 2.1$?

A. $3.35 \div 2.15$ B. $3.45 \div 2.15$ C. $3.35 \div 2.05$ D. $3.45 \div 2.05$

3.4 lies between 3.35 and 3.45.
2.1 lies between 2.05 and 2.15.

For the least possible value we require (the smaller of the limits for 3.4) ÷ (the larger of the limits for 2.1),

so we require $3.35 \div 2.15$.

So A gives the least possible value.

Exercise

In each of the following, numbers are given correct to 2 significant figures.

9 Which gives the **least** possible value to the quotient $4.3 \div 1.7$?

   A. $4.25 \div 1.75$ B. $4.25 \div 1.65$ C. $4.35 \div 1.75$ D. $4.35 \div 1.65$

10 Which gives the **least** possible value for the quotient $0.36 \div 2.3$?

   A. $0.355 \div 2.25$ B. $0.365 \div 2.25$ C. $0.355 \div 2.35$ D. $0.365 \div 2.35$

11 Which gives the **greatest** possible value for the quotient $5.4 \div 3.2$?

   A. $5.35 \div 3.15$ B. $5.35 \div 3.25$ C. $5.45 \div 3.25$ D. $5.45 \div 3.15$

12 Which gives the **greatest** possible value for the quotient $0.12 \div 8.6$?

   A. $0.125 \div 8.55$ B. $0.115 \div 8.55$ C. $0.125 \div 8.65$ D. $0.115 \div 8.65$

Continue with Section P

P Progress check

Exercise

1 The length and breadth of an envelope are measured to the nearest centimetre and found to be 22 cm and 11 cm. Find the possible range of the perimeter of the envelope.

2 The length of a square is 4.5 cm to the nearest tenth of a centimetre. Find the possible range of the perimeter of the square.

3 A piece of wood 0.4 m in length is cut off a larger piece which measures 3.2 m. What is the largest possible length of the remaining piece?

4 (a) A metal strip 5 cm long is cut from a piece 34 cm long. What are the limits of the length left?
   (b) If two such strips are cut off, what are the limits of the remaining length?

5 Find the limits between which the areas of the following must lie.
   (a) A rectangle of length 7 cm and breadth 4 cm.
   (b) A square of side 5 m.

6 If each of the numbers 5.6 and 4.7 is correct to 2 significant figures, which of the following gives the **least** possible value of the quotient $5.6 \div 4.7$?

   A. $5.55 \div 4.65$ B. $5.65 \div 4.75$ C. $5.55 \div 4.75$ D. $5.65 \div 4.65$

7 If each of the numbers 0.34 and 5.3 is correct to 2 significant figures, which of the following gives the **greatest** possible value of the quotient $0.34 \div 5.3$?

   A. $0.345 \div 5.35$ B. $0.345 \div 5.25$ C. $0.335 \div 5.35$ D. $0.335 \div 5.25$

Tell your teacher you have finished this unit

ANSWERS

UNIT I   Number Skills

A **1** $13+4+3=20$ **2** $20-6-4=10$ **3** $19-3+8=24$ **4** $30-15+5=20$
 5 $14+11-6=19$ **6** $27+17-16=28$ **7** 79 **8** 9 **9** 29 **10** 39
 11 11 **12** 1 **13** 0 **14** 3

B **1** 37 **2** 124 **3** 72.4 **4** 370 **5** 1240 **6** 724 **7** 3700 **8** 12 400
 9 7240 **10** 92 **11** 155.1 **12** 152.8 **13** 920 **14** 1551 **15** 1528
 16 9200 **17** 15 510 **18** 15 280 **19** 3 **20** 40 **21** 0.6 **22** 2.26
 23 1.34 **24** 0.51 **25** 0.226 **26** 0.134 **27** 0.051 **28** 0.0226
 29 0.0134 **30** 0.0051

C **1** 7.82 **2** 147.43 **3** 7.216 **4** 4.428 **5** 0.024 **6** 0.064 **7** 1.22
 8 348 **9** 0.024 **10** 6 **11** 230 **12** 23 **13** 32 **14** 35 **15** 28
 16 2000 **17** 400 **18** 20 **19** 1.04 **20** 3.45 **21** 2.32 **22** 1.6
 23 2.9 **24** 1.2 **25** 0.11 **26** 0.21 **27** 0.17

D **1** 33.02 **2** 33.77 **3** 8.8 **4** 4.89 **5** 13.78 **6** 11.7 **7** 4.7 **8** 2.5
 9 4.8 **10** 7.8

E **1** $\frac{3}{4}$ **2** $\frac{2}{3}$ **3** $\frac{9}{10}$ **4** $\frac{2}{3}$ **5** $\frac{2}{3}$ **6** $\frac{3}{4}$ **7** $\frac{30}{36}$ **8** $\frac{12}{32}$ **9** $\frac{24}{30}$ **10** $\frac{21}{49}$ **11** $\frac{8}{36}$
 12 $\frac{6}{75}$ **13** $\frac{3}{4}$ **14** $\frac{2}{5}$ **15** $\frac{7}{9}$ **16** $\frac{2}{5}$ **17** $\frac{7}{8}$ **18** $\frac{5}{6}$

F **1** $\frac{5}{12}$ **2** $\frac{2}{3}$ **3** $\frac{13}{21}$ **4** $1\frac{1}{2}$ **5** $1\frac{1}{6}$ **6** $1\frac{1}{15}$ **7** $\frac{1}{12}$ **8** $\frac{3}{8}$ **9** $\frac{4}{9}$ **10** $\frac{3}{10}$
 11 $\frac{13}{24}$ **12** $\frac{7}{15}$

G **1** 8 **2** 5 **3** 8 **4** 9 **5** 16 **6** 21 **7** 3 **8** 3 **9** 3 **10** 6 **11** 9
 12 15 **13** 3 **14** 4 **15** 4 **16** 9 **17** 12 **18** 20 **19** $\frac{1}{9}$ **20** $\frac{1}{8}$
 21 $\frac{1}{8}$ **22** $\frac{3}{8}$ **23** $\frac{1}{6}$ **24** $\frac{3}{25}$ **25** $\frac{2}{5}$ **26** $\frac{4}{15}$ **27** $\frac{3}{10}$

H Progress check (see page 251)

I **1** $4\frac{5}{6}$ **2** $4\frac{1}{3}$ **3** $1\frac{13}{18}$ **4** $1\frac{3}{4}$ **5** $\frac{37}{40}$ **6** $11\frac{19}{24}$ **7** $\frac{5}{12}$ **8** $3\frac{5}{8}$ **9** $2\frac{1}{6}$
 10 5 **11** $5\frac{1}{5}$ **12** 3 **13** $8\frac{1}{3}$ **14** $8\frac{1}{4}$ **15** 2

J **1** $1\frac{1}{3}$ **2** $\frac{4}{5}$ **3** $1\frac{1}{2}$ **4** $3\frac{1}{3}$ **5** $1\frac{5}{6}$ **6** $2\frac{1}{2}$

K **1** 0.5 **2** 0.8 **3** 0.15 **4** 0.04 **5** 0.02 **6** 0.75 **7** 0.88 **8** 0.22
 9 0.27 **10** 0.19 **11** 0.16 **12** 0.83

L **1** 50% **2** 80% **3** 15% **4** 4% **5** 2% **6** 75% **7** 37.5% **8** 44.4%
 9 18.2% **10** 31.3% **11** 21.9% **12** 33.3% **13** 0.2, 20%; 0.6, 60%;
 0.333...33.3%; 0.666...66.7% **14** 0.12, $\frac{3}{25}$ **15** 0.25, $\frac{1}{4}$ **16** 0.4, $\frac{2}{5}$
 17 0.08, $\frac{2}{25}$ **18** 0.125, $\frac{1}{8}$ **19** 0.075, $\frac{3}{40}$

M **1** 60 **2** 70 **3** 90 **4** 200 **5** 400 **6** 500 **7** 0.4 **8** 0.07 **9** 0.12
 10 0.002 **11** 0.013 **12** 0.14

N **1** E **2** E **3** C **4** D
O **1** A **2** B **3** C **4** B
P Progress check (see page 251)

UNIT 2   Holidays and Foreign Exchange

A **1** (a) £180 (b) £208 (c) £350 **2** (a) Between 6th July and 9th Sept.
 (b) £195 (c) £12.25 (d) £3 (e) From Luton; £5
B **1** December: £6.90 per hour   July: £5 per hour

C **1**

Age group	Total cost (£)
Under 12 years	306
12–13	628
14	1908
15	513
16–20	366
Paying adults	378
Total	4099

2

Age group	Total cost (£)
12–13	1670
14–15	1448
16–20	776
Paying adults	199
Total	4093

D **1** £196.50; greater by £13 **2** £233.50; greater by £13 **3** £2092.50
4 (a) £602.50 (b) £75 (c) £67.50 (d) £60; Total £805
E **1** 14775 pesetas **2** 9555 pesetas **3** 17966 pesetas **4** 16213 pesetas
5 £10.15 **6** £4.57 **7** £2.28 **8** £15.48 **9** £12.89
F **1** £295.65 francs **2** 511.37 francs **3** 911.04 francs **4** 1078.58 francs
5 £2.37 **6** £4.55 **7** £8.83 **8** £18.93 **9** £13.20
G **1** £1.87 **2** £2.73 **3** 4 roubles; £3.08 **4** 30 Kroner; £2.97
H **1** 37.50 dollars **2** 77525 lire **3** 985 pesetas **4** 53.55 DM
5 150 schillings
I Progress check (see page 251)
J **1** (a) Spain, Italy (b) U.S.A., Canada, France
3 177.80 dollars; 90.3 dollars **4** 2920 pesetas; 1020 pesetas
K **1** £185.12 **2** £33.44 **3** £274.56
L **1** £11.26 **2** £3.20 **3** £24.79 **4** £2.67 **5** £8.54 **6** £2.32 **7** £1.45
8 £3.66 **9** £3.61

M

Schillings	1	2	3	4	5	6	7	8	9
£	0.04	0.07	0.11	0.14	0.18	0.21	0.25	0.29	0.32

Schillings	10	20	30	40	50	60	70	80	90
£	0.36	0.71	1.07	1.43	1.79	2.14	2.50	2.86	3.21

N Progress check (see page 251)

UNIT 3 Rounding and Scientific Notation

A **1** Between 175 and 185 **2** Between 905 and 915 **3** Between 65 and 75
4 Between 595 and 605
B **1** Between 11 500 and 12 500 **2** Between 7750 and 7850
3 Between 62 500 000 and 63 500 000 **4** Between 55 000 and 65 000
5 Between 650 000 and 750 000 **6** Between 1950 and 2050
C **1** 8400 **2** 16 000 **3** 750 **4** 160 000 **5** 800 **6** 7000 **7** 8000
8 12 000 000
D **1** 14 000 **2** 4000 **3** 1100 **4** 31 000 **5** 14 000 g
6 14 480, 14 500, 14 000 **7** £79 200 **8** 12 300
E **1** 12 **2** 20 **3** 141 **4** 61 **5** 5 **6** 86

F **1** Between 41.55 and 41.65 **2** Between 5.665 and 5.675
3 Between 7.115 and 7.125 **4** Between 8.895 and 8.905
5 Between 3.1415 and 3.1425 **6** Between 7.5255 and 7.5265
G **1** 14.6 **2** 2.4 **3** 2.0 **4** 8.17 **5** 11.18 **6** 7.83 **7** 2.140 **8** 2.000
H **1** 0.2 m² **2** 2.20 m **3** 5.3 cm, 1.8 cm, 9.5 cm² **4** 0.8 m² **5** 16.7%
6 28.6% **7** 27.27%
I **1** $10^6, 10^7, 10^8$ **2** 10^3 **3** 10^6 **4** 10^9 **5** 10^8 **6** 10^{14} **7** 8.5×10^3
8 6.2×10^7 **9** 4.3×10^5 **10** 5.4×10^9 **11** 8.7×10^{13} **12** 3.2×10^8
13 1.23×10^5 **14** 4.2×10^7 **15** 6.75×10^8 **16** 8.76×10^4
17 5.08×10^5 **18** 9.46×10^3 **19** 1.76×10^{10} **20** 1.134×10^6 **21** 7.32×10^{33}
J $10^{-5}, 10^{-6}, 10^{-7}$ **2** 5×10^{-4} **3** 9.7×10^{-5} **4** 8.4×10^{-6}
5 6.2×10^{-4} **6** 7.52×10^{-7} **7** 8.03×10^{-10} **8** 5.24×10^{-3}
9 8.7×10^{-5} **10** 2.2×10^{-8} **11** 5.12×10^{-4} **12** 1.23×10^{-4}
13 6.8×10^{-5} **14** 5.26×10^{-3} **15** 6.18×10^{-6} **16** 5.789×10^{-7}
17 7.346×10^{-1}
K Progress check (see page 251)
L **1** 10^8 **2** 10^{10} **3** 10^{16} **4** 10^5 **5** 10^3 **6** 10^3 **7** 10^3 **8** 10^7
9 10^8 **10** 8×10^4 **11** 7.2×10^7 **12** 8×10^7 **13** 6.5×10^{15}
14 2×10^3 **15** 2.1×10 **16** 1.2×10^3 **17** 2.3×10^7 **18** 1.9×10^3
19 1.2×10^5
M **1** 10^2 **2** 10^{-5} **3** 10^{-4} **4** 10^{-6} **5** 10^{11} **6** 10^{-4} **7** 10^{17} **8** 10^{12}
9 10^{-3} **10** 8×10^{-10} **11** 6×10^{-4} **12** 9×10^3 **13** 8.2×10^{-17}
14 6.51×10^{-11} **15** 2.8×10^4 **16** 4×10^{-5} **17** 2×10^5 **18** 3×10^{-7}
19 1.1×10^{-11} **20** 4×10^8 **21** 2×10^{-6} **22** 1.2×10^{-7} **23** 1.5×10^6
24 1.8×10^{-14} **25** 1.44×10^{-6} **26** 1.9×10^{20} **27** 10^{-14} **28** 7×10^{-5}
29 6×10^{-6} **30** 7×10^{-3} **31** 4×10^8 **32** 9×10^{-5} **33** 4×10^2
N Progress check (see page 251)

UNIT 4 Spending Money

A **1** £0.69 **2** £2.02 **3** £2.07
B **1** £100.35 **2** £107.85 **3** £24.65
C **1** £139.20 **2** £17.10 **3** £21.12
D **1** £11.23 **2** £270 **3** £13.44
4 36p, 40p, £1.57, 63p, 50p
E **1** 123.6 m **2** 27 600 **3** 96 kg **4** £6804 **5** £48.60, £5.83
F **2** £63.92, £39.48, £35.16, £9.31, £10.43, £1.69
G **1** £50.83 **2** £40 **3** £44.18 **4** £12.58 **5** £16.84
H **1** £4424 **2** 506 000 kg **3** 13.2 knots **4** 55 words per minute **5** 87 kg
I **1** 2 star: 224200, 3 star: 177 000, 4 star: 778 800
2 Domestic: 219 million, Commercial: 56 million, Industrial: 73 million
J **3** 26%, 14%, 8%
K **1** 50–99: 25%, 100–149: 47.5%, 150 and over: 20%
L Progress check (see page 251)
M **1** (a) £80.50 (b) £7 (c) £6.13 **2** (a) £78 (b) £6.50 (c) £7.15
3 £5649, £538, £212.96 **4** £4611, £522, £170.38
N **1** (a) £128 (b) £171.50 **2** £193 **3** £5170 **4** £10 340

O **1** 80%, 43% 68%, 35%, 15%, 28% **2** 10%, 21%, 15%, 19%
3 (a) 25% (b) 23%
4 30%, 39%, 56%, 19%, 27%, 27%
P **1** profit: 33%, loss: 20%, loss: 14% **2** £160, 20% **3** 12% **4** 40p, 22%
5 £40, 22% **6** £13.80, 26% **7** 15% **8** 20%
Q **1** £2 **2** £32 **3** £100 **4** £10 000
R Progress check (see page 251)

UNIT 5 Probability

A **1** $\frac{1}{4}$ **2** $\frac{3}{4}$ **3** $\frac{1}{2}$ **4** $\frac{1}{2}$
B **1** $\frac{5}{6}$ **2** $\frac{1}{6}$ **3** $\frac{5}{6}$ **4** $\frac{1}{2}$ **5** $\frac{1}{3}$ **6** $\frac{1}{2}$ **7** $\frac{1}{3}$ **8** $\frac{2}{3}$
C **3** (a) $\frac{1}{26}$ (b) $\frac{1}{4}$ (c) $\frac{1}{26}$ (d) $\frac{1}{13}$ (e) $\frac{5}{52}$ (f) $\frac{11}{52}$ (g) $\frac{3}{13}$ (h) $\frac{9}{13}$
D **1** 0.25 **2** 0.5 **3** 1 **4** 0.75 **5** 0
E **1** 2000 **2** 500 **3** 250 **4** 1000 **5** 300 **6** 1000 **7** 2000 **8** 250
F **1** $\frac{1}{6}$ **2** $\frac{5}{12}$ **3** $\frac{5}{12}$
G **1** $\frac{1}{36}$ **2** $\frac{1}{36}$ **3** $\frac{5}{36}$ **4** $\frac{1}{2}$ **5** $\frac{5}{18}$ **6** $\frac{13}{18}$ **7** $\frac{1}{6}$ **8** $\frac{5}{6}$ **9** 1
H **3** $\frac{1}{4}$ **4** $\frac{3}{16}$ **5** $\frac{1}{4}$ **6** 0 **7** 1
I **2** $\frac{1}{12}$ **3** $\frac{1}{4}$ **4** $\frac{1}{3}$
J **2** (a) $\frac{1}{8}$ (b) $\frac{3}{8}$
K **1** $\frac{1}{52}$ **2** $\frac{1}{2}$ **3** $\frac{1}{6}$ **4** $\frac{1}{4}$ **5** $\frac{1}{104}$ **6** $\frac{1}{24}$ **7** $\frac{1}{208}$ **8** $\frac{1}{12}$ **9** $\frac{1}{36}$ **10** $\frac{1}{4}$
L **1** $\frac{6}{25}$ **2** $\frac{6}{25}$ **3** $\frac{4}{25}$ **4** (a) $\frac{4}{5}$ (b) $\frac{1}{5}$ (c) $\frac{4}{25}$
M **1** $\frac{1}{2}$ **2** $\frac{3}{10}$ **3** $\frac{1}{5}$ **4** $\frac{1}{4}$ **5** $\frac{3}{20}$ **6** $\frac{3}{50}$ **7** $\frac{1}{25}$ **8** $\frac{1}{10}$ **9** $\frac{3}{100}$ **10** $\frac{1}{8}$
11 $\frac{27}{1000}$ **12** $\frac{1}{125}$
N Progress check (see page 251)
O **1** (a) $\frac{1}{4}$ (b) $\frac{3}{4}$ **2** (a) $\frac{1}{10}$ (b) $\frac{9}{10}$ **3** (a) 0.06 (b) 0.94 **4** (a) $\frac{1}{5}$ (b) $\frac{4}{5}$
5 0.1
P **1** (a) $\frac{1}{8}$ (b) $\frac{1}{8}$ (c) $\frac{1}{8}$ **2** (a) $\frac{1}{216}$ (b) $\frac{25}{216}$ (c) $\frac{1}{216}$ **3** (a) $\frac{63}{1000}$ (b) $\frac{147}{1000}$
(c) $\frac{27}{1000}$ (d) $\frac{343}{1000}$ **4** (a) $\frac{1}{16}$ (b) $\frac{1}{16}$ (c) $\frac{1}{16}$ **5** (a) $\frac{1}{1296}$ (b) $\frac{25}{1296}$ (c) $\frac{1}{1296}$
6 (a) $\frac{1}{3}$ (b) $\frac{2}{3}$ (c) $\frac{1}{9}$ (d) $\frac{2}{9}$ (e) $\frac{4}{9}$ (f) $\frac{2}{9}$ (g) $\frac{1}{27}$ (h) $\frac{2}{27}$ (i) $\frac{4}{27}$ (j) $\frac{4}{27}$ (k) $\frac{4}{27}$
(l) $\frac{8}{27}$
Q **2** 6 **3** 6 **4** 6 **5** 24 **6** $\frac{1}{24}$
R Progress check (see page 251)

UNIT 6 Formulae

A **3** (a) Robert de Niro (Tues), Meryl Streep (Wed), Jeremy Irons (Sun), Ben Kingsley (Fri), Dustin Hoffman (Sun), Natasha Kinski (Tues), (b) Terry Wogan (Tues), Ronnie Corbett (Thur), Cilla Black (Thur), Lulu (Weds), David Bowie (Weds), (c) Duke of Edinburgh (Fri), Queen Elizabeth (Wed), Prince Charles (Sun), Princess Diana (Sat), Princess Anne (Tues), (d) Neil Kinnock (Sat), Margaret Thatcher (Tues), Ronald Reagan (Mon), (e) Tessa Sanderson (Weds), Daley Thompson (Weds), Glen Hoddle (Weds), Seve Ballesteros (Tues), Sebastian Coe (Sat).
B **1** 28 **2** 6 **3** 55
C **1** 84 m² **2** 78 m² **3** 5250 cm² **4** 450 m²
D **1** (a) 21.41 cm (b) 34.5 cm **2** (a) 356 m (b) 58.6 cm
3 (a) 94.2 cm (b) 37.68 cm (c) 25.12 cm **4** (a) 8.06 cm (b) 4.80 cm

E **1** (a) 56 m² (b) 12.42 cm² **2** (a) 78.5 cm² (b) 2826 m²
(c) 379.94 cm² ≈ 380 cm² **3** (a) 314 cm² (b) 314 cm² **4** (a) 216 cm²
(b) 2400 cm²

F **1** 15.54 m² **2** 62.13 cm² **3** 102.96 m² ≈ 103 m² **4** 9.42 cm²
5 480 mm² **6** 69 cm²

G **1** (a) 40 cm³ (b) 51 cm³ **2** (a) 512 cm³ (b) 15.625 cm³ ≈ 15.6 cm³
3 (a) 1256 cm³ (b) 1130.4 cm³ ≈ 1130 cm³ **4** 1 099 000 cm³

H **1** (a) 66 cm³ (b) 32 m³ **2** 500 cm³

I **1** 14 m **2** 10 m **3** 4 cm **4** 5 cm **5** 16 cm **6** 5.64 cm **7** 4.37 cm

J Progress check (see page 251).

K **1** (a) £1260 (b) £1680 **2** (a) £1380 (b) £1740 (c) £2288
3 (a) 13p (b) 18p (c) 22p (d) 38p (e) 68p **4** (a) £36.50 (b) £85.50
(c) £299

L **1** (a) 144 cm² (b) 33 cm² **2** (a) 201 cm² (b) 11 cm² **3** £1510
4 317 kg

M **1** 69 cm³ **2** 36 cm³ **3** 52 cm³ **4** (a) 47 cm³ (b) 27 cm³ (c) 33 cm³
5 (a) 113 m³ (b) 333 cm³ (c) 421 cm³ **6** 17 g

N **1** (a) 17 232 cm³ (b) 2462 cm³ (c) 20 000 cm³ **2** 7200 cm³
3 230 cm³ **4** 10 200 kg

O **1** 35 cm³ **2** 113 cm² **3** 880 g

P **1** (a) $d = \dfrac{c}{\pi}$ (b) $h = \dfrac{v}{lb}$ (c) $h = \dfrac{3v}{A}$

2 (a) $y = \dfrac{A}{3} - x$ (b) $b = \dfrac{A}{2h} - l$ (c) $h = \dfrac{A}{2\pi r} - r$

3 (a) $r = \sqrt{\dfrac{V}{\pi h}}$ (b) $r = \sqrt{\dfrac{A}{4\pi}}$ (c) $r = \sqrt{\dfrac{3V}{\pi h}}$

4 (a) $d = \dfrac{k^2}{V^2}$ (b) $a = \dfrac{b^2 - d^2}{4c}$ (c) $R = \dfrac{c^2 + h^2}{2h}$

5 $P = Tr^2$ **6** $d = \sqrt{\dfrac{V}{\sqrt{H}}}$ **7** $k = dt + m$

8 $r = \dfrac{100c}{Pn}$ **9** $x = \dfrac{r - q}{p}$ **10** $p = \dfrac{2}{q - 1}$

Q Progress check (see page 251)

UNIT 7   Mean, Median, and Mode

A **1** (a) 22, 23, 23, 24, 25, 26, 27 (b) 24
2 (a) 14 yr 4 mth, 14 yr 5 mth, 14 yr 6 mth, 14 yr 7 mth, 14 yr 8 mth,
14 yr 9 mth, 15 yr, 15 yr 1 mth, 15 yr 2 mth (b) 14 yr 8 mth
3 (a) 198 (b) 6 (c) 880 (d) 22 (e) 65
4 (a) 9 (b) 5.5 (c) 103 (d) 28.5 (e) 4

B **1** Median is 159 for group A and 157 for group B. Group A seems taller.
2 Median is 120 for group A and 122 for group B. Group B seems taller.

C **1** (a) 1 (b) 8 (c) 1 **2** (a) 4 (b) 14 (c) 5

D **1** (a) 8 (b) 17.5 (c) 6.4 (d) 176.5 (e) $2280 \div 40 = 57$ **2** 65
   3 44 years   **4** 83.6 kg
E **1** (a) 1.9 (b) Performed better on question in example.
   2 (a) 19 (b) 2.3 (c) 8.2 (d) 18.5
F **1** 4.74   **2** 1.85

G **1** $14\frac{1}{2}$ **3** (a)
   2 (a) 1 (b) 0 (c) 25

1	37
2	31
3	12
4	10
Total	90

(b) 1

H **1** (a) 7 (b) 32 (c) 25   **2** (a) 4 (b) 17 (c) 13   **3** (a) 21 (b) 29 (c) 8
   4 (a) 12.6 (b) 14.5 (c) 1.9   **5** (a) 28 (b) 88 (c) 60
I   **1** (a) 2   (c) 2 (d) 2 (e) the same
J   Progress check (see page 252)
K **1** (a) Frequencies: 2, 3, 5, 4, 8, 3, 1 (b) $82 - 23 = 59$ (c) $50 - 59$
   2 (a) Frequencies: 4, 18, 12, 9, 3 (b) $50 - 54$ (c) $64 - 40 = 24$
   3 (a) $79 - 42 = 37$ (b) Frequencies: 2, 4, 2, 9, 8, 4, 3, 4 (c) $60 - 64$
   4 (a) Frequencies: 3, 5, 8, 14, 20, 8 (b) $15 - 19$ (c) 29
L   **1** (b) $1335 \div 30 = 44.5$
   2 (a) $3139 \div 20 = 156.95 \approx 157$ (b) $558 \div 30 = 18.6$ (c) $705 \div 50 = 14.1$
   3 $615 \div 30 = 20.5$ **4** $1320 \div 60 = 22$
M **1** (a) Frequencies: 24, 26, 28, 22, 10, 8, 2 (b) 120 (c) 15 (d) 35%
   (e) (i) $\frac{28}{120} = \frac{7}{30}$ (ii) $\frac{50}{120} = \frac{5}{12}$
   2 (a) Frequencies: 3, 4, 5, 8, 3, 3, 3, 1 (b) 6 hours (c) 40% (d) $\frac{1}{3}$
   3 (a) 120 (b) $60 - 69$ (c) $40 - 49$ (d) $50 - 59$ (e) 89
   4 (a) $\frac{6}{36} = \frac{1}{6}$ (b) $\frac{3+2+1}{36} = \frac{1}{6}$ (c) 7 (d) $\frac{9}{36} = \frac{1}{4} = 25\%$
   5 (a) 4   (b) 24   (c) 1–3, 25–27, 28–30
N   Progress check (see page 252)

UNIT 8   Borrowing and Saving

A **1** £474.90   **2** £546.63
B **1** £25.20   **2** £510   **3** (a) £24 (b) £15.30 (c) £2.90 (d) £60 (e) £46.40
   4 £8720   **5** £10.80   **6** £38.95
   7 (a) £11.20 (b) £25 (c) £3.15 (d) £95 (e) £2.60   **8** £458.70
C **1** £83.20   **2** £126.10   **3** (a) £61.50 (b) £264.80   **4** £190.91
   5 £2784.60
D **1** £41.40   **2** £47.48
E **1** £36   **2** £180   **3** £204
F **1** £48.30   **2** £41.40   **3** £35.65   **4** £57.50   **5** £63.25   **6** £83.95
G **1** (a) £20 800 (b) £6 200   **2** (a) £17 000 (b) £3500   **3** (a) £16 740
   (b) £2260
   4 (a) £31 000 (b) £1000

H **1** (a) £124.35 (b) £29 844 (c) £14 844
　　2 (a) £136.80 (b) £41 040 (c) £23 040
　　3 (a) £191.00 (b) £34 380 (c) £14 380
　　4 (a) £171.00 (b) £51 300 (c) £28 800
I **1** £102.00 **2** £10 (minimum premium) **3** £33.60 **4** £138.00
J **1** £19.30 **2** £21.30 **3** £31.80 **4** £26.90 **5** £56.70 **6** £38.30
　　7 £65.40
K Progress check (see page 252)
L **1** £352.80 **2** £17 463.60 ≈ £17 460 **3** £2099.52 ≈ £2100
M **1** £1458 **2** £192 **3** £3584 **4** £34 969
N **1** £89.60 **2** £180.65
O **1** £66.39 **2** £276.55 **3** (a) £232 (b) £227.55 (c) less
P **1** £37.50 **2** £42 **3** £19.20 **4** £300 **5** £114.75
Q **1** (a) £720 (b) £44 (c) 6.1% **2** (a) £768 (b) £60 (c) 7.8%
　　3 357 shares **4** (a) 300 shares (b) £12 (c) 5.7%
　　5 (a) 10% (b) $12\frac{1}{2}$% So (b) provides greater yield. **6** £288
R **1** Property £1800, Building Society £1500, Shares £1300 **2** £1600
　　3 (a) Property (b) Building Society and Shares **4** 60%
S Progress check (see page 252)

UNIT 9　Pictorial Representation

A **1** Netherlands **2** Eire **3** greater **4** (a) 320 (b) 90 **5** D **6** No
　　8 Cars **9** Tractors **10** 30 **11** 125
B **1** (a) 400 (b) 820 (c) 350 **3** less **4** 7 **5** 8400 **6** 12 000
　　7 23 400 **8** Yes **9** No **10** Yes **11** No **12** increase **13** B
C **1** (a) 25% (b) 15% (c) 20% (d) Road and Air (e) No (f) £105 m
　　2 (a) Gas (b) 18%; 16% (c) 2906 million therms (d) North Sea Gas
　　3 (a) 0.14 mm (b) 29 (c) 200 (d) 47% (e) 4%
D **1** (a) April (b) December (c) (i) 6.8 (ii) 8.8 (d) 1.7 (e) North
　　2 (a) January (b) £2100 (c) £6000 (d) 35%
E **1** (a) 150 (b) 100 (c) 50 (d) 50 000 hours
　　(e) 225 000 hours, 250 000 hours, 175 000 hours (f) 700 000 hours
　　(g) 400 (h) 1750 hours
　　2 (c) 50 mm (f) 90 mm
F **1** (a) 3000 (b) 6000 (c) 1000 **2** (a) 25% (b) £80 000
　　(c) Materials used, Wages and salaries
　　3 (a) U.S.A. (b) West Germany (c) 400 000 (e) $\frac{1}{4}$
G **1** 120° **2** 90°
H Progress check (see page 252)
J **1** (a) October 1981 (b) February 1982
　　(c) February–June (d) 1980
　　(e) always increasing; school leavers
　　2 (a) 3.6 m (b) 2 m (c) 8.20 am (d) 9.50 am till 12.20 pm
　　3 (a) 21 g (b) B (c) 30 ℃ (d) 30 g (e) 23 g (f) 84 ℃
　　4 (a) 1979 and 1980 (b) £224 (c) £1.80; 45p (d) 12%

5 (a) 22 200 km (b) 14 600 (c) 7600 km (d) June (e) 630 (f) 23 months
6 Volume in ml per beat: 80; 100; 100; 100
7 (a) (i) 2 million (ii) 8.8 million (b) 3000 and 25 000
(c) 1931 (d) 1953
8 (a) 250 km and 450 km (b) 5 litres (c) 50 litres (d) 10 litres per 100 km
(e) greater
9 (a) 1800 hours (b) 1500 (c) 75%
K   Progress check (see page 252)

UNIT I0   Rates and Taxes

A   **1**   (a) £77.64 (b) £89.28 (c) £144.30 (d) £23.44   **2** £25.83, £89.99
3   £69.34, £146.66   **4**   £250.14, £441.36   **5**   £245.35, £594.65
6   (a) Incorrect: £127.85 (b) Correct (c) Incorrect: £122.50
B   **1**   £200.70   **2**   £104.33
C   **1**   (a) £22.40 (b) £105.28   **2** £156.80   **3** £107.20   **4** £155.64   **5** £35.34
D   **1**   £805   **2**   (a) £1004 (b) £1012 (c) £1511.52   **3**   (a) £1440 (b) £1884
E   **1**   (a) (i) £12.13 (ii) £10.82 (iii) £13.21 (iv) £12.62 (v) £11.59
(b) (i) £14.08 (ii) £12.57 (iii) £15.34 (iv) £14.66 (v) £13.45
2   £36.65   **3**   £504.40   **4**   £39.27
F   **1**   £4650   **2**   £2300   **3**   £3400
G   **1**   (a) £3400 (b) £8120   **2**   (a) £3420 (b) £2170   **3**   £10 010
H   **1**   (a) £2466   **2**   £1170   **3**   £765   **4**   £3532.50   **5**   £1701.90
6   (a) £1341.60 (b) £1983.60 (c) £1656 (d) £4152
7   (a) £1102.20 (b) £1857 (c) £2970.60 (d) £3621
I   **1**   (a) £5650 (b) £10 550 (c) £3165
2   (a) £10 140 (b) £3400 (c) £6740 (d)£2022
3   (a) £10 296 (b) £3550 (c) £6746 (d) £2023.8
J   Progress check (see page 252)
K   **1**   £1500
2   £6810.4   **3**   £6780
L   **1**   (a) £27.60   (b) £66.01   (c) £112.24   (d) £664.70   **2**   (a) £8   (b) £350
(c) £193   (d) £1211
M   **1**   (a) £256 (b) £286 (c) £387.60   **2**   £5203.80   **3**   £52.36   **4**   £495.80
less
5   £7.50 more
N   **1**   70p in the £   **2**   85p in the £   **3**   £1.20 in the £   **4**   £1.74 in the £

O

	Rate in the £	Surplus
1	89p	£30 000
2	£1.23	£300 000
3	96p	£16 000
4	£1.39	£3 400 000

P   Progress check (see page 252)

UNIT II   Index Numbers

B **1** 1983 = 100, 1984 = 110, 1985 = 127, 1986 = 138
   2 1983 = 100, 1984 = 110, 1985 = 125, 1986 = 150
C **1** 1983 = 100, 1984 = 112, 1985 = 120, 1986 = 136
   2 1982 = 100, 1983 = 128, 1984 = 160, 1985 = 188
   3 1981 = 100, 1982 = 110, 1983 = 125, 1984 = 145
D **1** 1950 = 100, 1955 = 102, 1960 = 104, 1965 = 108, 1970 = 110
   2 1979 = 100, 1980 = 108, 1981 = 116, 1982 = 132, 1983 = 148
   3 1960 = 100, 1965 = 107, 1970 = 98, 1975 = 85, 1980 = 70
   4 1979 = 100, 1980 = 125, 1981 = 150, 1982 = 160, 1983 = 165
E **1** £38.76 **2** £29 **3** £119 **4** (a) £335 (b) £350 **5** (a) £546
      (b) £572 **6** (a) £17600 (b) £19200
F **1** 1980 = 100, 1981 = 110, 1982 = 124, 1983 = 130, 1984 = 136
   2 (a) 1983 = 100, 1984 = 125, 1985 = 140
      (b) 1978 = 100, 1979 = 120, 1980 = 150
      (c) 1980 = 100, 1981 = 115, 1982 = 120
      (d) 1982 = 100, 1983 = 90, 1984 = 75
G **1** (a) 1980 = 100, 1981 = 115, 1982 = 120, 1983 = 130, 1984 = 135
   (c) 1981   (d) 1980 and 1983   (e) 15%   (f) (i) £78 (ii) £80.60
   (iii) £2.60   (g) £20.40
   2 1980 = 100, 1981 = 105, 1982 = 110, 1983 = 115
   3 1981 = 100, 1982 = 104, 1983 = 94, 1984 = 84, 1985 = 82
   4 1981 = 100, 1982 = 140, 1983 = 150, 1984 = 165, 1985 = 190
   (c) 1983 (d) 200p
H Progress check (see page 252)
I **1** (a) 1974 = 84, 1979 = 92, 1981 = 100, 1983 = 108
   (b) 1980 = 90, 1981 = 100, 1983 = 105, 1985 = 115
   2 1979 = 85, 1980 = 100, 1981 = 92.5, 1982 = 110, 1983 = 102.5
   3 1973 = 60, 1976 = 100, 1979 = 200, 1983 = 360, 1985 = 400
J **1** £105.60 **2** £3420 **3** 180 million tonnes **4** £3.00 worse off
K **1** (a) £97 (b) £97.80 (c) £86.40
   2 Weighted average height 1.5 metres
   3 Charlie's average mark is 58.75. Charlie will get the job.
   4 First C=57.4%; second B=55.6%, third A=53.9%
   5 First B = 59.6%; second A = 59.1%; third C = 53.4%
      French and Geography
L **1** Weighted index for 1972 = $\frac{2001}{20}$ = 100.05 $\approx$ 100
   2 Weighted index = 112.8 $\approx$ 113
   3 (a) Weights: bread 20; meat 37; potatoes 20; coffee 6; sugar 17.
   Total = 100.
      (b) Weighted price index = 152.19 $\approx$ 152 (c) 52%
   4 City household weighted average index = 100
      Country household weighted average index = 107
      County household expenditure has increased.
M Progress check (see page 252).

UNIT 12   Estimates and Errors

A **1** 715–725 **2** 5150–5250 **3** 1750–1850 **4** 13 500–14 500
5 4500–5500 **6** 350–450

B **1** 40 **2** 80 **3** 90 **4** 20 **5** 70 **6** 300 **7** 2000 **8** 70 **9** 700
10 100 **11** Wrong **12** Wrong **13** Possible **14** Possible **15** Wrong
16 Wrong

C **1** 150 **2** 980 **3** 2460 **4** 5720 **5** 1100 **6** 5200 **7** 2000
8 690 000 **9** 50 000 **10** 7200 **11** 2440 **12** 82 000 **13** 30 **14** 120
15 40

D **1** 18 **2** 163 **3** 5 **4** 20 **5** 5 **6** 18

E **1** 0.7 **2** 0.04 **3** 0.2 **4** 0.4 **5** wrong **6** wrong **7** possible
8 possible **9** wrong **10** wrong **11** possible **12** wrong

F **1** 0.3 **2** 0.071 **3** 0.802 **4** 4 **5** 4.3 **6** 4.26 **7** 1.4 **8** 0.0540
9 27 **10** 27.3 **11** 0.3 **12** 0.21

G **1** 4960 **2** 27 000 **3** 260 **4** 410 000 **5** 700 **6** 8 **7** 8.3 **8** 0.43
9 0.0184

H **1** C **2** D **3** C **4** A **5** C **6** D **7** E

I **1** 7.5 cm, 8.5 cm, 0.5 cm **2** 74.5 cm, 75.5 cm, 0.5 cm
3 12.5 kg, 13.5 kg, 0.5 kg **4** 5.5 l, 6.5 l, 0.5 l **5** 4.25 g, 4.35 g, 0.05 g
6 15.65 cm, 15.75 cm, 0.05 cm **7** 9.55 kg, 9.65 kg, 0.05 kg
8 48.25 km, 48.35 km, 0.05 km **9** 2.235 l, 2.245 l, 0.005 l
10 18.15 cm, 18.25 cm, 0.05 cm **11** 1.025 cm, 1.035 cm, 0.005 cm
12 123.5 m, 124.5 m, 0.5 m

J **1** 29.40 cm–29.80 cm **2** 35 900 kg–36 300 kg **3** 245 m–255 m
4 53.4 cm–54.6 cm **5** 450–550 **6** 381 g–387 g

K Progress check (see page 252)

L **1** 11.5 cm, 12.5 cm; 7.5 m, 8.5 m; 38.0 m, 42.0 m; 2.0 m
2 (a) 10.5 cm, 13.5 cm, 1.5 cm (b) 12.85 cm, 13.15 cm, 0.15 cm

M **1** 12.0 cm, 14.0 cm **2** 35.0 kg, 37.0 kg **3** (a) 34.0 cm, 36.0 cm
3 (b) 23.5 cm, 26.5 cm **4** smallest: 1.90 m, largest: 2.10 m

N **1** 8.75 m², 15.75 m² **2** 6.25 m², 12.25 m² **3** 11 25 cm², 21.25 cm²

O **1** (a) A (b) C **2** (a) B (b) E **3** (a) B (b) E **4** (a) A (b) C
5 (a) B (b) D **6** (a) E (b) A **7** (a) D (b) E **8** (a) E (b) C **9** A
10 C **11** D **12** A

P Progress check (see page 252)

Answers to Progress Checks

Teachers may want to cut this page from the book.

Unit 1
H **1** 17 **2** 20 **3** 3 **4** 7 **5** (a) 137 (b) 1370 (c) 13 700
6 (a) 83.2 (b) 832 (c) 8320 **7** (a) 2.43 (b) 0.243 (c) 0.0243 **8** 0.027
9 5.024 **10** 3.496 **11** 2.4 **12** 27 **13** 6.5 **14** 18.4 **15** 13.3
16 (a) $\frac{3}{5}$ (b) $\frac{3}{4}$ (c) $\frac{1}{2}$ **17** (a) $\frac{12}{18}$ (b) $\frac{8}{10}$ (c) $\frac{25}{30}$ **18** $\frac{4}{5}$ **19** (a) $\frac{1}{2}$ (b) $1\frac{1}{35}$
20 (a) $\frac{1}{2}$ (b) $\frac{3}{10}$ **21** (a) 9 (b) 12 (c) 9 **22** (a) $\frac{1}{32}$ (b) $\frac{2}{15}$ (c) $\frac{3}{8}$
P **1** $3\frac{14}{15}$ **2** $1\frac{5}{6}$ **3** $2\frac{11}{12}$ **4** $5\frac{1}{4}$ **5** $4\frac{8}{15}$ **6** (a) 0.6 (b) 0.3 (c) 0.25
7 (a) 25% (b) 40% (c) 70% **8** (a) 0.14 (b) 0.11 (c) 0.67
9 (a) 62.5% (b) 55.6% (c) 18.8% **10** (a) $\frac{1}{12}$ (b) $\frac{3}{50}$ (c) $\frac{3}{20}$
11 (a) 80 (b) 0.11 **12** B **13** D **14** A **15** B

Unit 2
I **1** £218 **2** £3768 **3** 2955 Pesetas **4** £18.58 **5** £5.39 **6** 85.68 Dm
N **1** (a) £1.49 (b) £3.20 (c) £1.44 **2** 1250 Pesetas **3** £185.50

Unit 3
K **1** Between 15 550 and 15 650 **2** Between 0.175 and 0.185
3 (a) 583 270 (b) 583 300 (c) 583 000 **4** 14 cm² **5** 15.1 cm **6** 27.84
7 28.247 **8** (a) 5.2×10^7 (b) 7.2×10^6 (c) 5.0×10^{12} (d) 6.3×10^{-5}
(e) 5.2×10^{-6} (f) 8.73×10^{-3} **9** (a) 5.64×10^6 (b) 4.23×10^{-3}
(c) 7.78×10^{-3}
N **1** 8×10^7 **2** 6.9×10^{-7} **3** 2.8×10^3 **4** 2.7×10^{-4} **5** 1.2×10^{13}
6 1.92×10^6 **7** 5×10^2 **8** 9×10^{-7}

Unit 4
L **1** (a) £44.40 (b) £7.40 **2** £115.20 **3** £64.60 **4** (a) 84
(b) 36 **5** £1.88
R **1** (a) £414 (b) £28.80 (c) £16.05 **2** £166.40 **3** 22% **4** 25%
5 63% **6** £8 **7** £150

Unit 5
N **1** (a) $\frac{1}{6}$ (b) $\frac{1}{2}$ (c) $\frac{1}{3}$ **2** $\frac{1}{3}$ **3** (a) $\frac{1}{4}$ (b) $\frac{1}{12}$ (c) $\frac{1}{6}$ (d) $\frac{3}{4}$
5 (a) 500 (b) 1000 (c) 2000 **6** (a) $\frac{1}{6}$ (b) $\frac{1}{3}$ (c) $\frac{1}{3}$
7 (a) $\frac{1}{36}$ (b) $\frac{1}{2}$ (c) $\frac{1}{12}$ (d) $\frac{1}{6}$ (e) $\frac{11}{36}$ **8** (a) $\frac{1}{36}$ (b) $\frac{1}{18}$ (c) $\frac{1}{6}$ (d) $\frac{1}{9}$
R **1** (a) $\frac{1}{5}$ (b) $\frac{4}{5}$ **2** $\frac{3}{16}$ **3** (a) $\frac{5}{36}$ (b) $\frac{5}{36}$ (c) $\frac{25}{36}$
4 789,798,879,897,978,987; Pr(879) = $\frac{1}{6}$

Unit 6
J **1** 150 m² **2** 50.24 cm **3** 7.78 cm **4** (a) 251.2 cm² (b) 628 cm³
5 (a) 16.79 cm² (b) 2.73 m² **6** 9.6 cm **7** 2.7 m
Q **1** (a) £1120 (b) £3648 (c) £6912 **2** 693 cm³ **3** 2240 kg **4** £282.60
5 (a) 28.26 m² (b) 1.57 m² (c) 18.84 m³
6 (a) $l = \dfrac{p - \pi d}{2}$ (b) $d = 63.7$ m

Unit 7

J **1** Set A: 18, range = 24; Set B: 25, range = 8
2 (a) 3 (b) frequencies: 3,11,15,18,4,3,2 (c) 2   **3** 8.5
N **1** (a) 72 (b) 14.5 (c) 60   **2** $450 \div 30 = 15$
3 (a) $6\frac{1}{2}$ (b) $\frac{18}{40} = 45\%$ (c) $\frac{24}{40} = \frac{3}{5}$
4 (a) 10–19 (b) 30–39 (c) 99 (d) $\frac{62}{200} = 31\%$ (e) 34

Unit 8

K **1** £33.75   **2** £36   **3** £252.50   **4** (a) £18.70 (b) £7200, £1300 (c) £71.28
5 £15   **6** £16.20
S **1** £9917.60 ≈ £9920   **2** £121.72   **3** (a) £272 (b) £20.40 (c) 7.5%

Unit 9

H **1** 72°   **2** (a) 10 (b) 30 (c) 86 (d) 20 (e) 20% (f) $\frac{1}{5}$ (g) 33
3 (a) £6022 m. (b) Motoring (c) 20% (d) Motoring and Clothing
K **1** (a) Bus (b) 1968 (c) 5600 million (d) 1977
2 (a) 36 minutes (b) 4 km/h (c) 7.5 km (d) 4.10 pm
3 (a) Sterling Area, Soviet Union and Eastern Europe
(b) Sterling Area, North America, Soviet Union (c) 20%
(d) £1040 m. (e) 5% (f) Sterling Area (g) £1288 m.

Unit 10

J **1** (a) £71.32 (b) £34.24 (c) £34.86 (d) £4.59   **2** (a) £98.04 (b) £291.96
3 £52.80   **4** £338   **5** (a) £2085 (b) £1645 (c) £575.75
P **1** £158.10   **2** £54.60   **3** 87p in the £; £14 000   **4** (a) £847 (b) £1363
(c) £477.05

Unit 11

H **1** 1971 = 100, 1972 = 110, 1973 = 112, 1974 = 115
2 1974 = 100, 1975 = 116, 1976 = 132; 1975 = 100, 1976 = 112, 1977 = 124
3 £18   **4** £1692   **5** (a) 1971 = 100, 1972 = 95, 1973 = 90, 1974 = 95
M **1** (a) 1975 = 76, 1976 = 100, 1977 = 116 (b) 1976 = 72, 1977 = 84,
1978 = 100   **2** £47.50   **3** Wage index 10%; industrial production index 9%;
wage index increased more.   **4** 93

Unit 12

K **1** 16 550–16 650
2 (a) 400 (b) 40 (c) 1000 (d) 360 (e) 1100 (f) 410 (g) 1060 (h) 2390
(i) 1080   **3** (a) 2325, 2300 (b) 212,210
4 (a) 0.08 (b) 0.04 (c) 0.2 (d) 0.32 (e) 0.091 (f) 29 (g) 28.8 (h) 7.32
(i) 0.0326   **5** (a) 7.05, 7.1 (b) 44.238, 44.2   **6** C   **7** E
8 (a) 3.15 g, 3.25 g, 0.05 g (b) 4.335 m, 4.345 m, 0.005 m
(c) 6.5 kg, 7.5 kg, 0.5 kg (d) 7.395 cm, 7.405 cm, 0.005 cm
9 10.20 cm and 10.60 cm   **10** 12.75 cm and 13.05 cm
P **1** 64 cm, 68 cm   **2** 17.80 cm, 18.20 cm   **3** 2.90 m
4 (a) 28.0 cm, 30.0 cm (b) 22.5 cm, 25.5 cm
5 (a) 22.75 cm², 33.75 cm² (b) 20.25 m², 30.25 m²   **6** C   **7** B

Printed in Great Britain by Richard Clay (The Chaucer Press) Ltd, Bungay, Suffolk